QUANTUM DOTS

SELECTED TOPICS IN ELECTRONICS AND SYSTEMS

Editor-in-Chief: **M. S. Shur**

Published

Vol. 6: Low Power VLSI Design and Technology
eds. *G. Yeap and F. Najm*

Vol. 7: Current Trends in Optical Amplifiers and Their Applications
ed. *T. P. Lee*

Vol. 8: Current Research and Developments in Optical Fiber Communications in China
eds. *Q.-M. Wang and T. P. Lee*

Vol. 9: Signal Compression: Coding of Speech, Audio, Text, Image and Video
ed. *N. Jayant*

Vol. 10: Emerging Optoelectronic Technologies and Applications
ed. *Y.-H. Lo*

Vol. 11: High Speed Semiconductor Lasers
ed. *S. A. Gurevich*

Vol. 12: Current Research on Optical Materials, Devices and Systems in Taiwan
eds. *S. Chi and T. P. Lee*

Vol. 13: High Speed Circuits for Lightwave Communications
ed. *K.-C. Wang*

Vol. 14: Quantum-Based Electronics and Devices
eds. *M. Dutta and M. A. Stroscio*

Vol. 15: Silicon and Beyond
eds. *M. S. Shur and T. A. Fjeldly*

Vol. 16: Advances in Semiconductor Lasers and Applications to Optoelectronics
eds. *M. Dutta and M. A. Stroscio*

Vol. 17: Frontiers in Electronics: From Materials to Systems
eds. *Y. S. Park, S. Luryi, M. S. Shur, J. M. Xu and A. Zaslavsky*

Vol. 18: Sensitive Skin
eds. *V. Lumelsky, M. S. Shur and S. Wagner*

Vol. 19: Advances in Surface Acoustic Wave Technology, Systems and Applications (Two volumes), volume 1
eds. *C. C. W. Ruppel and T. A. Fjeldly*

Vol. 20: Advances in Surface Acoustic Wave Technology, Systems and Applications (Two volumes), volume 2
eds. *C. C. W. Ruppel and T. A. Fjeldly*

Vol. 21: High Speed Integrated Circuit Technology, Towards 100 GHz Logic
ed. *M. Rodwell*

Vol. 22: Topics in High Field Transport in Semiconductors
eds. *K. F. Brennan and P. P. Ruden*

Vol. 23: Oxide Reliability: A Summary of Silicon Oxide Wearout, Breakdown, and Reliability
ed. *D. J. Dumin*

Vol. 24: CMOS RF Modeling, Characterization and Applications
eds. *M. J. Deen and T. A. Fjeldly*

Selected Topics in Electronics and Systems — Vol. 25

QUANTUM DOTS

Editors

E. Borovitskaya

Michael S. Shur

Rensselaer Polytechnic Institute, USA

World Scientific
New Jersey • London • Singapore • Hong Kong

Published by
World Scientific Publishing Co. Pte. Ltd.
P O Box 128, Farrer Road, Singapore 912805
USA office: Suite 1B, 1060 Main Street, River Edge, NJ 07661
UK office: 57 Shelton Street, Covent Garden, London WC2H 9HE

British Library Cataloguing-in-Publication Data
A catalogue record for this book is available from the British Library.

QUANTUM DOTS

Copyright © 2002 by World Scientific Publishing Co. Pte. Ltd.

All rights reserved. This book, or parts thereof, may not be reproduced in any form or by any means, electronic or mechanical, including photocopying, recording or any information storage and retrieval system now known or to be invented, without written permission from the Publisher.

For photocopying of material in this volume, please pay a copying fee through the Copyright Clearance Center, Inc., 222 Rosewood Drive, Danvers, MA 01923, USA. In this case permission to photocopy is not required from the publisher.

ISBN 981-02-4918-7

Printed in Singapore by Mainland Press

Solidity is an imperfect state
Within the cracked and dislocated Rear
Non-stoichiometric crystals dominate.
Stray Atoms sully and precipitate.

John Updike

CONTENTS

Low-Dimensional Systems   1
    E. Borovitskaya and M. S. Shur

Energy States in Quantum Dots   15
    A. J. Williamson

Self-Organized Quantum Dots   45
    A. R. Woll, P. Rugheimer, and M. G. Lagally

Growth, Structures, and Optical Properties of III-Nitride Quantum Dots   79
    D. Huang, M. A. Reshchikov, and H. Morkoç

Theory of Threshold Characteristics of Quantum Dot Lasers: Effects of
Quantum Dot Parameter Dispersion   111
    L. V. Asryan and R. A. Suris

Applications of Quantum Dots in Semiconductor Lasers   177
    N. N. Ledentsov, V. M. Ustinov, D. Bimberg, J. A. Lott, and
    Zh. I. Alferov

LOW-DIMENSIONAL SYSTEMS

ELENA BOROVITSKAYA and MICHAEL S. SHUR
*Department of Electrical, Computer, and Systems Engineering,
Rensselaer Polytechnic Institute, Troy, NY 12180-3590, USA*

In 1932, H. P. Rocksby [1] discovered that the red or yellow color of some silicate glasses could be linked to microscopic inclusions of CdSe and CdS. It was not until 1985 when these changes in color were linked to the energy states determined by quantum confinement in these CdS or CdSe "quantum dots" [2]. Since that time, the number of experimental studies of quantum dots has literally exploded, stimulated, in particular, by the development of mature Molecular Beam Epitaxy technology, which rapidly became the technology of choice for growing quantum dots in different materials systems. Fig. 1 clearly illustrates the increased interest in quantum dots.

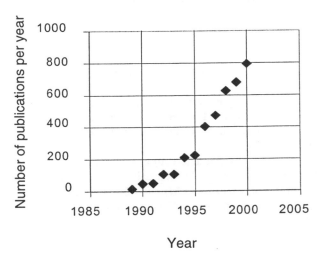

Fig. 1. Increase of publications on quantum dots.

More recently, a rapid progress in nanofabrication techniques allowed researchers to use lateral structures fabricated using different lithography techniques in order to create artificial quantum dots. Example of such technology is x-ray nanolithography, e-beam lithography, and nanoprinting.

Fig. 2 shows the predicted evolution of the minimum semiconductor device feature size according to the International Technology Roadmap for Semiconductors [3]

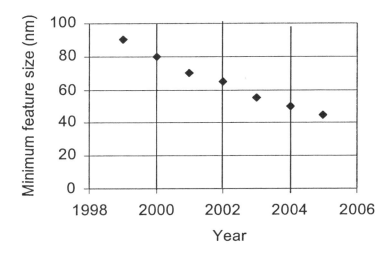

Fig. 2. Minimum feature size.

As can be seen from the figure, even the mainstream semiconductor device feature size (for transistors like those used in personal computers or cell phones) is projected to reach the dimension comparable with that of the largest quantum dots (on the order of 30 to 50 nm).

At the same time, modern characterization techniques have evolved to the point that individual atoms can be routinely seen using Transmission Electron Microscopy (TEM), Atomic Force Microscopy (AFM) and other methods. As an example, Fig. 3 shows nanometer size crystallites in a CdS film deposited from a solution on a flexible substrate (view foil).

1 nm

Fig. 3. Nanometer size crystallite in a CdS film deposited on view foil. One can distinguish individual atoms. [4]

Quantum dots, with sizes ranging from a few hundreds to many thousands of atoms, have already found applications, primarily in nonlinear optics and in the Q-switching of lasers. A quantum dot laser is yet to enter the mainstream but might soon emerge as the key application for quantum dot materials. Another potential application involves quantum computing. However, the full impact of quantum dot technology is still to come.

Many people refer to quantum dots as "artificial atoms". This comparison highlights two properties of a quantum dot: a relatively small number of electrons in the dot and many body effects by which the properties of the dot could be dramatically changed by adding just one electron. We can extend this analogy further by saying that two or more quantum dots might form an "artificial molecule". An array of quantum dots can form an artificial two-dimensional crystal. All in all, this points out the way to creating new artificially engineered quantum dot materials.

The idea of using quantum effects in thin layers of materials has been discussed since late 50'[5]. In his famous paper of 1962, Keldysh [6]developed the theory of the electron motion in a crystal with superimposed periodic potential. In 1970, Esaki and Tsu propose to use superlattices for the observation of the negative differential resistance. [7] In 1971, Alferov et al. reported on the $GaP_{0.3}As_{0.7}$/GaAs superlattice. [8] Also, in 1971, Kazarinov and Suris[9] proposed the idea of a unipolar laser using quantized subbands in thin semiconductor quantum wells. (It took nearly 25 years before this idea finally found its implementation in a quantum cascade laser. [10] The development of AlGaAs/GaAs materials system, with nearly perfect match in lattice constants stimulated the further development of these pioneering ideas, and studies of superlattices and quantum wells quickly became a mainstream of semiconductor research.

A quantum well might be implemented as a thin layer of semiconductor with a given energy gap is sandwiched between two slabs of another semiconductors with higher energy gap or it might form at a heterointerface due to a surface electric field, see Fig. 4.

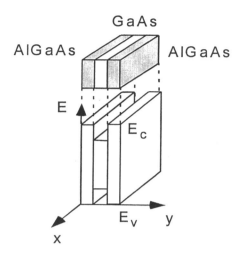

Fig. 4. AlGaAs/GaAs quantum well.

The thickness of the narrow gap semiconductor must be smaller or at least on order of the carrier mean free path between collisions with impurities of phonons. Hence, the quantum well thickness usually ranges from a fraction of a nanometer to a 10 or 20 nm. This thickness should be also smaller than the Broglie wavelength

$$\lambda = \frac{h}{p}, \tag{1}$$

where $p = m_n v$ is the electron momentum, m_n is the electron effective mass, and v is the electron velocity. Assuming that $v \sim v_{th}$, where

$$v_{th} = \sqrt{\frac{3k_B T}{m_n}},$$

is the thermal velocity, k_B is the Boltzmann constant, and T is temperature, we obtain

$$\lambda = \frac{6.22 \text{ nm}}{\sqrt{\frac{m_n}{m_o} \frac{T}{300}}} \tag{2}$$

where $m_o = 9.11 \times 10^{-31}$ kg is the electron mass. For example, for GaAs, $m/m_o = 0.067$, and, at $T = 300$ K, $v_{th} \sim 4.5 \times 10^5$ m/s, and $\lambda = 24$ nm. The electron mean free path, λ_{mean} can be estimated as

$$\lambda_{mean} = v_{th}\tau, \tag{3}$$

where the momentum relaxation time

$$\tau = \frac{m_n \mu}{q}, \tag{4}$$

q is the electronic charge, and μ is the low field mobility. For GaAs, at $T = 300$ K, the mobility could be as high as 8500 cm^2/V-s, and, hence, $\tau \sim 3.2 \; 10^{-13}$ s, and $\lambda_{mean} \sim 146$ nm. These estimates set up scales for observation of quantum effects in low dimensional structures.

Still another scale involves the energy level separation in a quantum well, where the motion of carriers in the direction perpendicular to the heterointerfaces is quantized, meaning that this motion involves discrete (quantum) energy levels.

The lowest energy levels for a square potential well can be estimated as follows:

$$E_j - E_c = \frac{\pi^2 \hbar^2}{2 m_n d^2} j^2. \tag{5}$$

Here j is the quantum number labeling the levels, and d is the thickness of the quantum well. For the quantization to be important, the difference between the levels should be much larger then the thermal energy $k_B T$, that is,

$$\frac{\pi^2 \hbar^2}{2 m_n d^2} \gg k_B T. \tag{6}$$

Using this condition, we find, for example, that in GaAs where m_n/m_0 0.067, the levels are quantized at room temperature when d 150 Å.

In the direction parallel to the heterointerfaces, the electronic motion is not restricted. Hence, the wave function for a two-dimensional electron gas can be presented as

$$\psi = f(y) \exp(i k_x x + i k_z z) \tag{7}$$

where $f(y)$ may be approximated by the wave function obtained from the solution of a one-dimensional Schroedinger equation for the infinitely deep quantum well:

$$f(y) = \left(\frac{2}{d}\right)^{1/2} \sin\left(\frac{\pi j}{d} y\right) \tag{8}$$

The term $\exp(ik_x x + ik_z z)$ in the wave function describing the electronic motion in directions x and z is similar to that of free electrons. This is understandable since electrons move freely in these directions. The dependence of the electron energy on the wave vector for a two-dimensional electron gas is given by

$$E - E_y = \frac{\hbar^2 (k_x^2 + k_z^2)}{2m_n}. \tag{9}$$

The k_y-component is absent in Eq. (9), since the motion in the y-direction is quantized. Each quantum level, E_j, given by Eq. (8) corresponds to an energy subband described by Eq. (8); see Fig. 5.

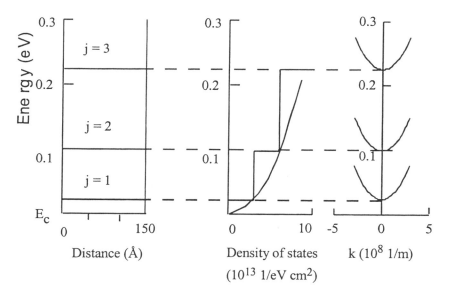

Fig. 5. Energy levels (bottoms of subbands), density of states for quantum wellstructure, and energy versus $k = (k_x^2 + k_z^2)^{1/2}$ for two-dimensional electron gas in GaAs quantum well.

Electrons in a quantum well form a two-dimensional electron gas (2DEG), and the 2DEG properties have been studied extensively since 1970's leading to discoveries of the integer [11] and fractional [12] Hall effect. In 1974 Dingle et al. [13] demonstrated the step-like absorption spectrum determined by two-dimensional density-of-states in quantum wells. Since 1980s, High Electron Mobility Transistors have emerged as superior high-speed devices using 2DEG, implemented first using AlGaAs/GaAs materials system [14] and, more recently, using AlGaN/GaN [15] or AlInGaN/InGaN materials systems. [16] In 1980's, researchers started intensive investigations of quantum wires [17, 18] (see Fig 6), and possible applications of quantum wires for lasers and other optoelectronic devices have been suggested. [19, 20]

This evolution from a three dimensional crystal to a quantum well and to a quantum wire has dramatic consequences for the energy spectra and for the electron density of states (see Fig. 6).

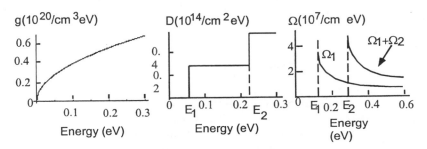

Fig. 6. Densities of states versus energy for three–dimensional (g), two–dimensional (D), and one–dimensional (Ω) electron gases in GaAs conduction band. Only the two lowest subbands are accounted for two–dimensional and one–dimensional electron gases. (From [21])

Fig. 7 illustrates the transition from one-dimensional quantum wire to a "zero dimensional" quantum box, i.e. a quantum dot.

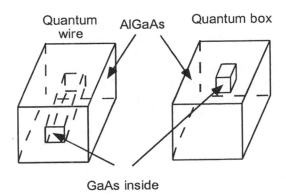

Fig. 7. Quantum wire and quantum box. (From [21])

In case of a three-dimensional confinement, when the particle is "locked" in rectangular potential box, and the electron potential $U(x)=0$ at $0<x<a$, $0<y<b$, $0<z<c$, and $U_0 \to \infty$ elsewhere, the energy levels are given by

$$E_n = \frac{\pi^2 \hbar^2}{2m} \left(\frac{n_1^2}{a^2} + \frac{n_2^2}{b^2} + \frac{n_3^2}{c^2} \right), \qquad n_1, n_2, n_3 = 1,2,3. \qquad (10)$$

The energy of ground state can be estimated as $E_0 \sim \hbar^2/ml^2$, where l is the characteristic linear size of system. In a quantum dot of a small potential depth the quantum level doesn't exist. The level appears if the potential depth exceed:

$$U_{0\min} = \frac{\pi^2 \hbar^2}{8ml^2} . \qquad (11)$$

However, this simple approach does not fully account for the physics of quantum wells, quantum wires and quantum dots. The space dependence of the effective mass in heterostructures might lead to a strong interdependence of the longitudinal and transverse motion in systems with heterointerfaces, such as quantum wells [22], quantum wires [23], and quantum dots. [24] For quantum wells and quantum wires, such interdependence results in additional non-parabolicity of the longitudinal motion in quantum wells. At large energies, comparable to or exceeding the depth of a quantum well, the effective mass changes sign and the longitudinal two-dimensional spectrum terminates at a certain critical value of the longitudinal momentum. The space dependence of the effective mass also strongly increases the transmission through a potential barrier for electrons with large incidence angles. These effects should have important consequences for impact ionization and tunneling phenomena in heterostructures. In particular, in many cases, the impact ionization electric field is limited by the cladding layers, and not by the material of the quantum well.

MOCVD grown quantum dots usually have pyramidal shape. [25] For such geometry, the Schroedinger equation in the effective mass approximation has to be solved numerically. [26] This is another important difference between the reality and a simple model of a quantum dot described by Eq. (10).

As mentioned above, the first realization of nano-size semiconductor inclusions into the glass was reported in 1932. In 1995, Goldstein et al demonstrated self-organized quantum dots in InAs/GaAs materials system. This was followed by the demonstration of quantum dots in a variety of other systems, such as InGaAs/GaAs [27], InGaAs/AlGaAs [28], InAlAs/AlGaAs [29,30], InP/InGaP [31,32] grown on GaAs substrates as well as InAs/InGaAs [33], InAs/InAlAs [34], InAs/InP [35] grown on InP substrates, and GaInP/GaP [36], InAs/GaP [37], InP/GaP [37] grown on GaP substrates. In 1990, Eaglesham and Cerullo [38] demonstrated that the quantum dot islands of Ge on Si formed in the Stranski-Krastanov (SK) [39] process of growth. This coherent growth was explained in terms of elastic deformation around of the islands, which partially compensated for the lattice mismatch.

Many books and review articles deal with the physics, technology, and applications of quantum dot structures.

In 1995, Richard Turton gave a popular review of emerging technology of "tiny" 0-dimensional devices" (quantum dots). [40]

In their book published in 1997, L. Jacak et al [41] reviewed the theory of quantum dots and experiments on quantum dot systems with typical dot sizes of 20 to 30 nm. They also included a short review of quantum dots growth methods. The book included the description of the quantum dot properties in a magnetic fields and the theory of exciton condensation and many-electron Quantum dots.

The book [42] by Bimberg et al (1998) gives a brief history of the subject and links the quantum dot properties to possible device applications. They reviewed various techniques of the quantum dot growth, including the concept of self-organization. The book also includes a detailed review of the theory of quantum dots, including the results of the numerical modeling of electrical and optical properties of quantum dots. The book also covers quantum dot based photonic devices.

The review paper of Randall et al [43] summarized the advances in microfabrication technology and discussed the possible use of structures with three dimensional electron confinement.

Smith [44] called the quantum dots the fourth generation of semiconductor technology. Harman [45] considered mesoscopical phenomena in quantum dots. Kastner [46] considered the problem of a current flow through the system of artificial atoms. Jefferson and Hausler [47] discussed similarities and differences between electronic properties of quantum dots (artificial atoms) and real atoms. They showed that the electron correlation phenomena in quantum dots could lead to new physical effects, which are absent in real atoms. The properties of electron transport in the system with quantum dots were also discussed in [48]. Alivisatos [49] gave a brief review of the optical and electrical properties of semiconductor clusters, nanocrystals, and quantum dots. Schweizer and Griesinger [50]. discussed the fabrication technology of quantum dot structures. Woggon and Gaponenko [51] reviewed excitons in quantum dots.

More recent reviews [52, 53, 54] surveyed the properties of self-organized quantum dots and the properties of quantum dot based injection lasers.

The infrared absorption properties of quantum dots were considered in Reference [55]. The optical properties of quantum crystallites with special emphasis on the Stark effect and nonlinear optical phenomena were discussed in Reference [56]. References [57, and 58] reviewed linear and nonlinear optical properties of II-VI semiconductor quantum dots in a glass matrix.

In 1982, Arakawa and Sakaki [59] proposed to use quantum dots for injection lasers. Such lasers have become one of the most interesting applications of quantum dot technology. [60, 61, 62, 63, 64, 65, 66] Other possible applications include storage devices, infrared detectors, and quantum computing.

Leading experts in quantum dot theory and technology contributed to this book that gives a comprehensive review of all aspects of quantum dot systems.

In the first chapter, Williamson discusses energy states in quantum dots, including the effects of strain and many body effects. The second chapter by Woll et al reviews the concepts of self-assembly and self-ordering of quantum dots in semiconductor systems.

In Chapter 3, Huang et al. discuss growth, structures, and optical properties of III-nitride quantum dots. Finally, Chapters 4 and 5 deal with quantum dot lasers.

Paraphrasing Enrico Fermi, we could say that if you were confused about quantum dots before reading this book, you might be still confused after reading it. But on a higher level.

---

[1] H. P. Rocksby, J. Soc. Glass Tech. **16**, 171, 1932.

[2] A.I. Ekimov, Al. L. Efros and A.A. Onushchenko, "Quantum size effect in semiconductor microcrystals", Sol. State Comm. **56** (1985) 921–924.

[3] http://public.itrs.net/

[4] Courtesy of BITs, Inc.

[5] V.N. Lutskii, "Quantum size effect-present state and perspectives of experimental investigations" Physica Status Solidi A, **1** (1970) 199 -- 200.

[6] L.V. Keldysh, Fiz. Tverd.Tela, 4 2236 (1962) [Sov. Phys. –Sol. State] 4 1658 (1963)].

[7] L. Esaki, R. Tsu, "Superlattice and negative differential conductivity in semiconductors", IBM Journal of Research and Development 14 (1970) 61--65.

[8] Zh. I. Alferov, Yu. V. Zhilyaev, Yu. V. Shmartsev, "Splitting of the conduction band in a 'superlattice' based on GaP_xAs_{1-x}", Fizika i Tekhnika Poluprovodnikov, [Sov.Phys. Semicond.] **5**, (1971), 196 --198.

[9] R. F Kazarinov, R. A. Suris, "Possibility of amplification of electromagnetic waves in a semiconductor with a superlattice", Fizika i Tekhnika Poluprovodnikov, [Sov.Phys. Semicond] **5**, (1971) 797--800.

[10] J. Faist, F. Capasso, D.L. Sivco, A.L. Hutchinson, C. Sirtori, S.N.G. Chu, A.Y. Cho, "Quantum cascade laser: temperature dependence of the performance characteristics and high T_0 operation", Appl. Phys. Lett. **65** (1994) 2901--2903.

[11] K. von Klitzing, G. Dorda, M. Pepper, "New method for high-accuracy determination of the fine-structure constant based on quantized Hall resistance" Phys. Rev. Lett. **45**, (1980), 494--497.

[12] D.C. Tsui, H. L. Stormer, A. C. Gossard, "Two-dimensional magnetotransport in the extreme quantum limit", Phys. Rev. Lett., **48**, (1982) 1559 --1562.

[13] R. Dingle, W. Wiegmann and C. H. Henry, "Quantum states of confined carriers in very thin $Al_xGa_{1-x}As$-GaAs-$Al_xGa_{1-x}As$ heterostructures", Phys. Rev. Lett, **33**, (1974) 827 -- 830.

[14] T. Mimura, S. Hiyamizu, T. Fujii, and K. Nambu, Jpn. J. Appl. Phys. 19, (1980) L225-227

[15] M. A. Khan, J. M. Van Hove, J. N. Kuznia, and D. T. Olson, Appl. Phys. Lett. 58, (1991) 2408

[16] M. S. Shur, B. Gelmont, C. Saavedra-Munoz, and G. Kelner, Potential of Wide Band Gap Semiconductor Devices for High Temperature Applications, Invited, in Proceedings of 5th Conference "Silicon Carbide and Related Compounds", Institute of

Physics Conference Series Number 137, Institute of Physics Publishing, M. G. Spencer, R. P. Devaty, J. A. Edmond, M. A. Khan, R. Kaplan, and M. Rahman, Eds. Bristol and Philadelphia (1994), pp. 691-694

[17] E. Kapon, M. C. Tamargo and D. M. Hwang, "Molecular beam epitaxy of GaAs/AlGaAs superlattice heterostructures on nonplanar substrates", Appl. Phys. Lett. **50**, (1987) 347 --349.

[18] M. Sweeny, J. Xu, M. S. Shur, "Hole Subbands in One-Dimensional Quantum Well Wires", Superlattices and Microstructures, **4** (1988) 623--626.

[19] J. Xu, M. S. Shur, and M. Sweeny, "Electronic and Optoelectronic Laser Devices Utilizing Light Hole Properties", United States Patent, # 4,999,682, March 12 (1991).

[20] J. Xu, M. S. Shur, and M. Sweeny, "Electronic and optoelectronic devices utilizing light hole properties", United States Patent, #4,899,201, February 6 (1990).

[21] M. S. Shur "Introduction to Electronic Devices", Wiley, New York, 1996.

[22] M. Dyakonov and M. S. Shur, "Consequences of Space Dependence of Effective Mass in Heterostructures" , J. Appl. Phys. **84** (1998) 3726--3730.

[23] E. Borovitskaya and M. S. Shur, "Consequences of Space Dependence of effective Mass in Quantum Wires", Solid State Electronics, **44** (2000) 1293 -- 1296 .

[24] E. Borovitskaya and M. S. Shur, "Consequences of Space Dependence of effective Mass in Quantum Dots" Solid State Electronics, **44** (2000) 1609 -- 1612.

[25] J. Oshinowo, M. Nishioka, S. Ishida and Y. Arakawa," Highly uniform InGaAs/GaAs quantum dots (approximately 15 nm) by metalorganic chemical vapor deposition" Appl. Phys. Lett., **65** (1994) 1421--1423.

[26] M. Grundmann, O.Stier and D.Bimberg, "InAs/GaAs pyramidal quantum dots: strain distribution, optical phonons, and electronic structure" Phys.Rev. B, **52**, (1995) 11969 -- 11981.

[27] D.Leonard, M.Krishnamurthy, L.M.Reaves, S.P. Den Baars, and P.M. Petroff, "Direct formation of quantum-sized dots from uniform coherent islands of InGaAs on GaAs surfaces", Appl. Phys. Lett. **63** (1993) 3203 -- 3205.

[28] A.E. Zhukov, A.Yu. Egorov, A.R. Kovsh, V.M. Ustinov, N.N. Ledentsov, M.V. Maximov, A.F. Tsatsul'nikov, S.V. Zaitsev, N.Yu Gordeev, P.S. Kop'ev, D. Bimberg, and Zh.I. Alferov, "Injection heterolaser based on an array of vertically aligned InGaAs quantum dots in an AlGaAs matrix", Semiconductors **31** (1997) 411-- 414.

[29] R. Leon, S. Fafard, D. Leonard, J.L. Merz, and P.M. Petroff, "Visible luminescence from semiconductor quantum dots in large ensembles", Appl. Phys. Lett. **67** (1995) 521 -- 523.

[30] A.F. Tsatsul'nikov, A.Yu. Egorov, P.S. Kop'ev, A.R. Kovsh, N.N. Ledentsov, M.V. Maximov, A.A. Suvorova, V.M. Ustinov, B.V. Volovik, A.E. Zhukov, M. Grundmann, D.Bimberg, and Zh.I. Alferov, "Optical properties of InAlAs quantum dots in an AlGaAs matrix", Appl. Surf. Sci. **123/124** (1998) 381 -- 384 .

[31] N. Carlsson, W. Seifert, A. Petersson, P. Castrillo, M.E. Pistol, and L. Samuelson "Study if the two-dimensional-three-dimensional growth mode transition in metal-organic vapor phase epitaxy of GaInP/InP quantum-sized structures", Appl. Phys. Lett. **65** (1994) 3093 -- 3095.

[32] M.K. Zundel, P. Specht, K. Eberl, N. Y. Jin-Phillipp, and F. Phillipp. "Structural and optical properties of vertically aligned InP quantum dots", Appl. Phys. Lett, **71** (1997) 2972 -- 2974.

[33] V.M. Ustinov, E. R. Weber, S. Ruvimov, Z. Liliental-Weber, A.E. Zhukov, A.Yu. Egorov, A.R. Kovsh, A.F. Tsatsul'nikov, and P.S. Kop'ev, "Effect of matrix on InAs self-orqanized quantum dots on InP substrate", Appl. Phys. Lett. **72** (1998) 362 -- 364.

[34] S. Fafard, Z. Wasilewski, J. Mc Caffrey, S. Raymond, and S. Charbonneau, "InAs self-assembled quantum dots on InP by molecular beam epitaxy", Appl. Phys. Lett. **68** (1996) 991 -- 993.

[35] A. Ponchet, A. Le Corre, H. L'Haridon, B. Lambert, and S. Salaun, "Relationship between self-organization and size of InAs islands on InP(001) grown by gas-source molecular beam epitaxy", App. Phys. Lett. **67** (1995) 1850 -- 1852.

[36] J.-W. Lee, A.T. Schremer, D. Fekete, and J.M. Ballantyne, "GaInP/GaP partially ordered layer type-I strained quantum well", Appl. Phys. Lett. **69** (1997) 4236 -- 4238.

[37] B. Junno, T. Junno, M.S. Miller, and L. Samuelson, "A reflection high-energy electron diffraction and atomic force microscopy study of the chemical beam epitaxial growth of InAs and InP islands on (001) GaP", Appl. Phys. Lett. **72** (1998) 954 -- 956.

[38] D.J. Eaglesham and M.Cerullo, "Dislocation-free Stranski-Krastanow growth of Ge on Si(100)" Phys.Rev.Lett., **64** (1990) 1943-1946.

[39] I.N.Stranski and Von L. Krastanow, Akad.Wiss. Lit. Mainz Math.-Natur. Kl. Iib **146**, (1939) 797.

[40] R. Turton. "The Quantum Dot: a Journey into the Future of Microelectronic", Oxford Univ. Pr., 1996.

[41] L.Jacak, P.Hawrylak, A. Wojs "Quantum Dots", Springer Verlag, Berlin, 1998.

[42] D. Bimberg, M. Grundmann, N.N. Ledentsov "Quantum Dot Heterostructures", John Wiley & Son Ltd, New York, 1999.

[43] J.N. Randall, M. A. Reed, G. A. Frazier, "Nanoelectronics: fanciful physics or real devices?" J. Vac. Sci. Technol. B, **7** (1989) 1398--1404.

[44] T.P. Smith, III "Semiconductor science: the fourth generation" Proceedings of the IEEE, **79** (1991) 1181--1187.

[45] K. Harmans, "Next electron, please...(mesoscopic physics)" Physics World, **5** (1992) 50--53.

[46] M. A. Kastner, "Artificial Atoms", Physics Today, **46** (1993) 24 --31.

[47] J. H. Jefferson, W. Hausler," Quantum dots and artificial atoms", Mol. Phys. Rep. (Poland), **17** (1997) 81 --103.

[48] C.G. Smith,"Low-dimensional quantum devices" Reports on Progress in Physics, **59** (1996) 235 -- 282.

[49] A. P. Alivisatos,"Semiconductors clusters, nanocrystals, and quantum dots", Science, **271** (1996) 933 --937.

[50] H. Schweizer, U.A.Griesinger, V. Harle, F. Adler, M. Burkard, F. Barth, J. Hommel, C. Kaden, J. Kovac, J. Kuhn, B. Klepser, G. Lehr, F. Prins, F. Scholz, M.H. Pilkuhn, J. Straka, A. Forchel, G.W. Smith, "Optoelectronics nanostructures: Physics and Technology", Proc. SPIE – Int.Soc.Opt.Eng., **2399** (1995) 407 --432.

[51] U. Woggon, S. V. Gaponenko, "Excitons in Quantum Dots" Phys.St.Solidi B, **189** (1995) 285 --343.
[52] N. N. Ledentsov, V. M. Ustinov, V. A. Shchukin, P. S. Kop'ev, Zh. I. Alferov, D. Bimberg, "Quantum dot heterostructures: fabrication, properties, lasers." Semiconductors (Sov.Phys.) **32**, (1998) 343--365.
[53] M. Grundmann, F. Heinrichdorff, N.N. Ledentsov, D. Bimberg, "New semiconductor lasers based on quantum dots" Laser and Optoelectronik, **30** (1998) 70--77.
[54] N.N. Ledentsov, M. Grundmann, F. Heinrichsdorff, D. Bimberg, V.M. Ustinov, A.E. Zhukov, M.V. Maximov, Zh.I. Alferov, J.A. Lott, "Quantum-dot heterostructure lasers" IEEE J. of Selected Topics in Quantum Electronics, **6**, (2000) 439--451.
[55] D. Heitmann, J.P. Kotthaus, "The spectroscopy of quantum dot arrays", Physics Today, **46** (1993) 56 --63.
[56] L.E. Brus, "Quantum crystallites and nonlinear optics", Appl. Phys. A, **A53** (1991) 465 -- 474.
[57] C. Klingshirn,"Linear and nonlinear optics of wide-gap II-VI semiconductors", Phys. St. Solidi B, **202** (1997) 857 --871.
[58] L.E. Brus, J.K. Trautman, "Nanocrystals and nano-optics", Philosophical Transactions of the Royal Society, Series A, **353** (1995) 313 -- 321.
[59] Y. Arakawa and H. Sakaki,"Multidimensional Quantum Well laser and temperature Dependence of Its Threshold Current", Appl. Phys. Lett. **40** (1982) 939--941
[60] J.-P. Leburton, F.H. Julien, Y. Lyanda-Geller,"Advances concepts in intersubband unipolar lasers", Int. Journal of High Speed Electronics and Systems, **9** (1998) 1163 -- 1188.
[61] J.H. Marsh, D. Bhattacharyya, A.S. Helmy, E.A. Avrutin, A.C. Bryce, "Engineering quantum-dot lasers" Physics E, **8** (2000) 154 --163.
[62] K. Eberl, "Quantum-dot lasers", Physics World, **10** (1997) 47 --50.
[63] S. Fafard, "Quantum dots promise a new dimension for semiconductor lasers" Photonics Spectra, **31** (1997) 160 --164.
[64] C. Weisbuch, "Physics of semiconductor lasers" Ann. Phys. (France), **20** (1995) 353-- 378.
[65] P. Bhattacharya, "Tunnel injection lasers" in M. Dutta, M. A. Stroscio "Advances in semiconductor lasers and applications to optoelectronics ", World Scientific, 2000, p. 1, p. 235.
[66] A.E. Zhukov, V.M. Ustinov, and Zh. I. Alferov "Device characteristics of low-threshold quantum-dot lasers" " in M. Dutta, M. A. Stroscio "Advances in semiconductor lasers and applications to optoelectronics ", World Scientific, 2000, p. 263.

Elena Borovitskaya received her M.S. degree in Physics from branch of Theoretical Physics, University of Nizhnii Novgord, Russia, Ph.D. in Physics and Mathematics from Applied Physics Institute, Nizhnii Novgorod. She has held research or faculty positions at different institutions, including Applied Physics Institute of Russian Academy of Science (RAS), Nizhnii Novgorod, Institute for Microstructures RAS, Nizhnii Novgorod, University of Nizhnii Novgorod, Institute for Solid States Physics RAS, Moscow, Technical University of Wuerzburg, Rensselaer Polytechnic Institute, Troy, NY.

Michael Shur received his M.S.E.E. degree (with honors) from St. Petersburg Electrotechnical Institute, Ph.D. in Physics and Mathematics from A. F. Ioffe Institute of Physics and Technology and Doctor of Science in Physics and Mathematics degree from A. F. Ioffe Institute in 1992. He has held research or faculty positions at different universities, including A. F. Ioffe Institute, Cornell, University of Minnesota, and University of Virginia, where he was John Money Professor of Electrical Engineering and served as Director of Applied Electrophysics Laboratories. He is now Patricia W. and C. Sheldon Roberts'48 Professor of Solid State Electronics, Professor of Physics, Applied Physics and Astronomy, Professor of Information Technology, and Acting Director of Center for Integrated Electronics and Electronics Manufacturing at RPI, NY.

Dr. Shur is Fellow of IEEE, Fellow of the American Physical Society, a member of Eta Kappa Nu, Sigma Xi, and Tau Beta Pi, Electrochemical Society; Electromagnetic Academy, Materials Research Society, SPIE, ASEE, and an elected member and former Chair of US Commission D, International Union of Radio Science. In 2001, he was elected Member-at-Large of USNC of URSI. Dr. Shur is Editor-in-Chief of the International Journal of High Speed Electronics and Systems, Editor of a book series on Selected Topics in Electronics and Systems published by World Scientific, and Member, Honorary Board of Solid State Electronics. In 1990–1993, he served as an Associate Editor of IEEE Transactions on Electron Devices. Dr. Shur has served as Chair, Program Chair, Technical Chair, Organizing and Program Committee Member of many IEEE conferences and as Member, IEEE Awards Committee. He is one of co-developers of AIM-Spice. In 1994, the Saint Petersburg State Technical University awarded him an Honorary Doctorate. He is also a co-author of the paper that received the best paper award at GOMAC-98 and a co-author of the best poster paper award at MRS-99. In 1999, he received van der Ziel Award from ISDRS-99 and Commendation for Excellence in Technical Communications from Laser Focus World. In 2000, he was listed by ISA as one of the most quoted researchers in his field.

WEB site: http://nina.ecse.rpi.edu/shur/

ENERGY STATES IN QUANTUM DOTS

ANDREW J. WILLIAMSON

*Lawrence Livermore National Labs., 7000 East Ave.,
Livermore, CA 94550, USA
williamson10@llnl.gov*

We describe a procedure for calculating the electronic structure of semiconductor quantum dots containing over one million atoms. The single particle electron levels are calculated by solving a Hamiltonian constructed from screened atomic pseudopotentials. Effects beyond the single particle level such as electron and hole exchange and correlation interactions are described using a configuration interaction (CI) approach. Application of these methods to the calculation of the optical absorption spectrum, Coulomb repulsions and multi-exciton binding energies of InGaAs self-assembled quantum dots are presented.

Keywords: Semiconductor quantum dots; electronic structure theory; empirical pseudopotentials; optical properties.

1. Introduction

As can be seen from the other chapters in this book, the past decade has witnessed a series of significant advances in the growth of semiconductor heterostructures and in particular, the growth of self-assembled or Stranski–Krastanow quantum dots. Even more dramatic than the advances in the growth of these structures has been the development of extremely sophisticated techniques for measuring the optical, electronic and transport properties of these quantum dot systems. The availability of such high quality measurements for a wide range of properties presents theorists with the formidable challenge of constructing models that can explain the origins of these experimental observations in terms of the underlying energy states of the quantum dots.

Calculating such energy states in self-assembled semiconductor quantum dots is made particularly challenging by a number of factors:

(1) The quantum dots contain a large number of atoms. A typical self-assembled quantum dot has a base of ~ 300 Å and a height of ~ 50 Å. Therefore the dot itself may contain $\sim 10^5$ atoms. This dot then needs to be surrounded by a barrier material to isolate it from other dots. A representative system containing both the dot and barrier therefore typically contains $\sim 10^6$ atoms.

(2) By the nature of the growth process, self-assembled quantum dots are highly strained. For example, in the most common InAs/GaAs material combination,

the lattice mismatch is 7%. Therefore, an accurate solution of the strain profile in the system is required before the electronic structure can be calculated.

(3) The valence band maximum in III–V semiconductor materials is three-fold degenerate (in the absence of strain and spin-orbit splitting), therefore any realistic approach must describe at least the band mixing between the three valence band edge states.

(4) In InAs the bulk band gap is 0.42 eV and the spin-orbit splitting is 0.38 eV, therefore the mixing between valence and conduction band states and split off states also have to be taken into account.

(5) In a zero-dimensional InAs/GaAs quantum dot system, the charge carriers are artificially confined inside the dot, which is typically smaller than the bulk excitonic radius — the "Strong Confinement" regime. This dramatically enhances the Coulomb interaction between charges in the dot and strongly modifies the dielectric screening.

The standard theoretical approach to this problem has been to adapt effective mass based techniques and their more sophisticated extension, the $\mathbf{k} \cdot \mathbf{p}$ method, which has been extremely successful in explaining a range of properties of bulk semiconductors. In essence, these methods expand the single particle wave functions of the system in a basis of bulk Bloch orbitals derived from the Brillouin zone center (Γ point). If a sufficient number of basis states are included, this expansion provides an excellent description of the band structure of the bulk material close to the Γ point. However, errors in the predicted energy rapidly increase as one moves away from the zone center. To improve the description of the band structure away from the zone center, one typically includes more and more basis functions in the expansion of the wave functions. The successes of this effective mass model in describing spectroscopic and transport properties in both three-dimensional bulk systems and two-dimensional quantum well structures are well documented. This model has also been demonstrated to be able to provide at least a qualitative picture of the energy states in zero dimensional systems. However, some of this success is mitigated by the fact that often the parameters in the model have to be refit to the nanostructure system itself. Recently, a direct comparison[1] between an 8-band $\mathbf{k} \cdot \mathbf{p}$ calculation and a full pseudopotential calculation for a pyramidal InAs quantum dot embedded within bulk GaAs showed a generally good agreement between the two techniques. However, the higher symmetry imposed by the $\mathbf{k} \cdot \mathbf{p}$ approach acts to omit certain energy level splittings and polarization anisotropies. The $\mathbf{k} \cdot \mathbf{p}$ approach also over confined both electrons and holes level by $\sim$50 meV.

In this article we instead describe our method of choice, the Empirical Pseudopotential Method (EPM) approach to calculating the energy states in semiconductor quantum dots. This approach has several advantages over the conventional effective mass approach to the problem:

(1) Once a pseudopotential has been developed to describe the bulk system, there are no adjustable parameters for describing a heterostructure system.
(2) The EPM method has the same accuracy, whether it is describing a three-dimensional bulk system, a two-dimensional quantum well, a one-dimensional quantum wire or a zero-dimensional quantum dot.
(3) The EPM method provides an *atomistic* description of the system. Therefore the correct symmetry of the underlying crystal lattice is reproduced. In addition, an atomistic description of interfaces between materials is also maintained.

We begin with a detailed description of each of the stages required in a typical EPM calculation of the energy states in a semiconductor quantum dot. Then we describe a series of applications of the EPM technique to studying a range of optical and electronic properties of self-assembled quantum dots.

2. Description of Pseudopotential Techniques

To calculate the energy states associated with various electronic excitations in self-assembled quantum dots requires three stages of calculation:

(i) *Assume the shape and composition and compute the strain*:

We first construct a supercell containing both the quantum dot and surrounding barrier material. Sufficient barrier material is used, so that when periodic boundary conditions are applied to the system, the electronic and strain interactions between dots in neighboring cells is negligible. The atomic positions within the supercell are relaxed by minimizing the strain energy described by an atomistic force field[2,3] including bond bending, bond stretching and bond bending-bond stretching interactions. More details of the atomistic relaxation are given in Sec. 2.1.

(ii) *Setup the pseudopotential single-particle equation*:

A single-particle Schrödinger equation is set up at the relaxed atomic positions, $\{\mathbf{R}_{n\alpha}\}$

$$\hat{H}\psi_i(\mathbf{r}) = \left\{ -\frac{\beta}{2}\nabla^2 + \sum_{n\alpha} \hat{v}_\alpha(\mathbf{r} - \mathbf{R}_{n\alpha}) \right\} \psi_i(\mathbf{r}) = \epsilon_i \psi_i(\mathbf{r}). \quad (1)$$

The potential for the system is written as a sum of strain-dependent, screened atomic pseudopotentials, v_α, that are fit to bulk properties extracted from experiment and first-principles calculations. For more details of the constructing of the Hamiltonian see Secs. 2.2 and 2.3.

(iii) *Calculate the screened, inter-particle many-body interactions*:

The calculated single particle wave functions are used to compute the electron–electron, electron–hole and hole–hole direct, J_{ee}, J_{eh}, J_{hh}, and exchange K_{ee}, K_{eh}, K_{hh} Coulomb energies. For more details see Sec. 2.5.

The main approximations involved in our method are: (a) the fit of the pseudopotential to the experimental data of bulk materials is never perfect (see Table 2) and (b) we neglect self-consistent iterations in that we assume that the screened pseudopotential drawn from a bulk calculation is appropriate for the dot. Our numerical convergence parameters are (i) the size of the GaAs barrier separating periodic images of the dots, and (ii) the number of basis functions used in the expansion of the wave functions (see Sec. 2.4 for more details).

2.1. *Calculation of the strain profile*

To obtain at *atomistic* description of the strain profile in a heterostructure system we construct an expression for the strain energy in terms of few-body potentials between actual atoms

$$E_{\text{strain}} = \sum_{ij} V_2(\mathbf{R}_i - \mathbf{R}_j) + \sum_{ijk} V_3(\hat{\Theta}_{ijk}) + \cdots, \quad (2)$$

where V_2 is a two-body term, V_3 is a three-body function of the bond angle, $\hat{\Theta}_{ijk}$. The functional form of these terms is taken to be strain-independent. The strain is determined by minimizing E_{strain} with respect to atomic positions $\{\mathbf{R}\}$.

Our chosen expression for the elastic strain energy, is a generalization (G-VFF) of the original Valence Force Field (VFF)[2,4] model. Our implementation of the VFF includes bond stretching, bond angle bending and bond-length/bond-angle interaction terms. This enables us to accurately reproduce the C_{11}, C_{12} and C_{44} elastic constants in a zincblende bulk material. We have also included higher order bond stretching terms, which yield the correct dependence of the Young's modulus with pressure. The expression for the G-VFF total energy is:

$$E_{\text{G-VFF}} = \sum_i \sum_j^{nn_i} \frac{3}{8} [\alpha_{ij}^{(1)} \Delta d_{ij}^2 + \alpha_{ij}^{(2)} \Delta d_{ij}^3]$$

$$+ \sum_i \sum_{k>j}^{nn_i} \frac{3\beta_{jik}}{8 d_{ij}^0 d_{ik}^0} [(\mathbf{R}_j - \mathbf{R}_i) \cdot (\mathbf{R}_k - \mathbf{R}_i) - \cos\theta_{jik}^0 d_{ij}^0 d_{ik}^0]^2$$

$$+ \sum_i \sum_{k>j}^{nn_i} \frac{3\sigma_{ijk}}{d_{ik}^0} \Delta d_{ij} [(\mathbf{R}_j - \mathbf{R}_i) \cdot (\mathbf{R}_k - \mathbf{R}_i) - \cos\theta_{jik}^0 d_{ij}^0 d_{ik}^0], \quad (3)$$

where $\Delta d_{ij} = [[(R_i - R_j)^2 - {d_{ij}^0}^2]/d_{ij}^0]$. Here $\mathbf{R}_i$ is the coordinate of atom i and d_{ij}^0 is the ideal (unrelaxed) bond distance between atom types of i and j. Also, θ_{jik}^0 is the ideal (unrelaxed) angle of the bond angle $j - i - k$. The $\sum^{nn_i}$ denotes summation over the nearest neighbors of atom i. The bond stretching, bond angle bending, and bond-length/bond-angle interaction coefficients $\alpha_{ij}^{(1)} (\equiv \alpha)$, β_{jik}, σ_{jik} are related to the elastic constants in a pure zincblende structure in the following way,

$$C_{11} + 2C_{12} = \frac{\sqrt{3}}{4d_0}(3\alpha + \beta - 6\sigma),$$

$$C_{11} - C_{12} = \frac{\sqrt{3}}{d_0}\beta, \qquad (4)$$

$$C_{44} = \frac{\sqrt{3}}{d_0}\frac{[(\alpha+\beta)(\alpha\beta-\sigma^2) - 2\sigma^3 + 2\alpha\beta\sigma]}{(\alpha+\beta+2\sigma)^2}.$$

The second-order bond stretching coefficient $\alpha^{(2)}$ is related to the pressure derivative of the Young's modulus by dB/dP, where $B = (C_{11} + 2C_{12})/3$ is the Young's modulus. The G-VFF parameters and the resulting elastic constants are shown in Table 1 for GaAs and InAs crystals.

For an InGaAs alloy system, the bond angle and bond-length/bond-angle interaction parameters β, σ for the mixed cation Ga–As–In bond-angle are taken as the algebraic average of the In–As–In and Ga–As–Ga values. In Fig. 1 we show the result of a G-VFF calculation for the strain profile in a lens shaped InAs quantum dot embedded within GaAs. The trace of the strain is plotted in a (010) plane through the center of the dot. It shows that, to a first approximation, the InAs dot is subjected to a uniform compressive strain and the GaAs barrier is slightly expanded around the interface with the dot.

Table 1. Input G-VFF parameters α, β, σ to Eq. (3) in units of 10^3 dyne/cm and their resulting elastic constants C_{11}, C_{12}, C_{44} in units of 10^{11} dyne/cm^2.

	α	β	σ	$\alpha^{(2)}$	C_{11}	C_{12}	C_{44}
GaAs	32.153	9.370	−4.099	−105	12.11	5.48	6.04
InAs	21.674	5.760	−5.753	−112	8.33	4.53	3.80

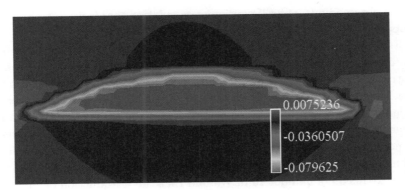

Fig. 1. Contour plot of the hydrostatic strain profile in a lens shaped, self-assembled InAs/GaAs quantum dot with a base of 252 Å and height of 35 Å.

2.2. Constructing the single particle Hamiltonian

We construct the single-particle Hamiltonian as:

$$\hat{H} = -\frac{\beta}{2}\nabla^2 + \sum_{n\alpha} \hat{v}_\alpha(\mathbf{r} - \mathbf{R}_{n\alpha}), \quad (5)$$

where $\mathbf{R}_{n\alpha}$ is the G-VFF relaxed position of the nth atom of type α. Here $\hat{v}_\alpha(\mathbf{r})$ is a screened empirical pseudopotential for atomic type α. It contains a local part and a nonlocal, spin-orbit interaction part.

The local potential part is designed to include dependence on the local hydrostatic strain $\text{Tr}(\epsilon)$:

$$v_\alpha^{\text{loc}}(r;\epsilon) = v_\alpha^{\text{eq}}(r;0)[1 + \gamma_\alpha \text{Tr}(\epsilon)], \quad (6)$$

where the γ_α is a fitting parameter. The zero strain potential $v_\alpha^{\text{eq}}(r;0)$ is expressed in reciprocal space q as:

$$v(q) = a_0(q^2 - a_1)/[a_2 e^{a_3 q^2} - 1], \quad (7)$$

where the $a_{0,1,2,3}$ are fitting parameters. The local hydrostatic strain $\text{Tr}(\epsilon)$ for a given atom at $\mathbf{R}$ is defined as $\Omega_R/\Omega_0 - 1$, where Ω_R is the volume of the tetrahedron formed by the four atoms bonded to the atom at $\mathbf{R}$. Ω_0 is the volume of that tetrahedron in the unstrained condition. The need for explicit dependence of the atomic pseudopotential on strain in Eq. (6) results from the following: While the description in Eq. (5) of the total pseudopotential as a superposition of atomic potentials situated at specific sites, $\{R_{n\alpha}\}$, does capture the correct local symmetries in the system, the absence of a self-consistent treatment of the Schrödinger equation deprives the potential from changing in response to strain. In the absence of a strain-dependent term, the volume dependence of the energy of the bulk valence band maximum is incorrect. While self-consistent descriptions show that the volume deformation potential $a_v = dE_v/d\ln\Omega$ of the valence band maximum is *negative* for GaAs, GaSb, InAs, InSb and for all II–VI this qualitative behavior cannot be obtained by a nonself-consistent calculation that lacks a strain dependent pseudopotential.

In Eq. (5), the kinetic energy of the electrons has been scaled by a factor of β. The origin of this term is as follows: In an accurate description of the crystal band structure, such as the GW method,[5] a general, spatially nonlocal potential, $V(r,r')$, is needed to describe the self-energy term. In the absence of such a term the occupied band width of an inhomogeneous electron gas is too large compared to the exact many-body result. To a first approximation, however, the leading effects of this nonlocal potential, $V(r,r')$, can be represented by scaling the kinetic energy. This can be seen by Fourier transforming $V(r,r')$ in reciprocal space, q, then making a Taylor expansion of q about zero. We find that the introduction of such a kinetic energy scaling, β permits a simultaneous fit of both the effective masses and energy gaps. In this study, we fit $\beta = 1.23$ for both GaAs and InAs.

Table 2. Fitted bulk electronic properties for GaAs and InAs using the screened atomic pseudopotentials, in Eq. (6). The hydrostatic deformation potential of the band gap and Γ_{15v} levels are denoted by a_{gap} and $a_{\Gamma_{15v}}$. The biaxial deformation potential is denoted by b and the spin-orbit splittings at the Γ_{15v} and L_{1v} points are denoted by Δ_0 and Δ_1.

Property	GaAs		InAs	
	EPM	Expt[7]	EPM	Expt[7]
E_{gap}	1.527	1.52	0.424	0.42
$E_{X_{5v}}$	−2.697	−2.96	−2.330	−2.40
$E_{X_{1c}}$	1.981	1.98	2.205	2.34
$E_{X_{3c}}$	2.52	2.50	2.719	2.54
$E_{L_{3v}}$	−1.01	−1.30	−5.76	−6.30
$E_{L_{1c}}$	2.36	1.81	1.668	1.71
m_e^*	0.066	0.067	0.024	0.023
$m_{hh}^*[100]$	0.342	0.40	0.385	0.35
$m_{hh}^*[111]$	0.866	0.57	0.994	0.85
$m_{lh}^*[100]$	0.093	0.082	0.030	0.026
a_{gap}	−7.88	−8.33	−6.79	−5.7
$a_{\Gamma_{15v}}$	−1.11	−1.0	−0.826	−1.0
b	−1.559	−1.7	−1.62	−1.7
Δ_0	0.34	0.34	0.36	0.38
Δ_1	0.177	0.22	0.26	0.27

The pseudopotential parameters in Eqs. (6) and (7) were fitted to the bulk band structures, experimental deformation potentials and effective masses and first-principles calculations of the valence band offsets of GaAs and InAs. The alloy bowing parameter for the GaInAs band gap (0.6 eV) is also fitted. The properties chosen for the fit and the quality of the fit for bulk InAs and bulk GaAs are given in Table 2. We see that unlike the LDA, here we accurately reproduce the bulk band gaps and the bulk effective masses. One significant difference in our parameter set, to that used in conventional $\mathbf{k} \cdot \mathbf{p}$ studies, is our choice of a negative magnitude for the valence band deformation potential, a_v, which we have obtained from LAPW calculations.[6]

2.3. *The inclusion of the spin-orbit interaction*

To calculate the spin-orbit interaction, each wave function must be represented by spin-up and spin-down components. The additional term in the Hamiltonian which describes the spin-orbit interaction is implemented in q space as a matrix (between plane wave k_1 and k_2). More specifically, we have:

$$\hat{H}_{SO} = \sum_l |l\rangle V_l^{SO}(\mathbf{r}) \mathbf{L} \cdot \mathbf{S} \langle l|. \tag{8}$$

Here $|l\rangle$ is the projection operator of spatial angular momentum l, $\mathbf{L}$ is the spatial angular momentum operator, $\mathbf{S}$ is the Dirac spin operator (matrix between spin up and down components), and $V_l^{SO}(\mathbf{r})$ is a potential representing the spin-orbit interaction due to relativistic effects of core electron states. In a plane wave basis $|k_1\rangle$, and $|k_2\rangle$, the $\hat{H}_{SO}$ can be rewritten as:

$$\langle k_1|\hat{H}_{SO}|k_2\rangle = \sum_l \frac{4i\pi}{\Omega}(2l+1)\left[\frac{dP_l(\cos\theta_{k_1k_2})}{d\cos\theta_{k_1k_2}}\right](\hat{k}_1 \times \hat{k}_2)\cdot\hat{S}$$

$$\times \int_0^\infty V_l^{SO}(r)j_l(|k_1|r)j_l(|k_2|r)r^2\,dr. \qquad (9)$$

Here, $\theta_{k_1k_2}$ is the angle between k_1 and k_2, Ω is the volume of the unit cell, j_l is the spherical Bessel function. In our calculations, we only include the effects of $l=1$ (p-states), and use a Gaussian model for $V_l^{SO}(r)$.

2.4. Solving the single particle Hamiltonian

We have developed two techniques for solving for the eigenstates of Eq. (5). The choice of technique depends on the size of the heterostructure system being studied, i.e., the number of electrons in the system, and the desired level of accuracy. The two techniques are distinguished by (i) the choice of the basis set for expanding the single particle eigenstates of Eq. (5), and (ii) the algorithm used to solve for the eigenstates within that particular basis. In our first approach we expand the eigenstates in a *plane wave* basis and use the Folded Spectrum Method to obtain eigenstates in a given energy window. In the second approach, we expand the eigenstates in a basis of *bulk Bloch orbitals* and use a Lanzcos algorithm to obtain eigenstates.

Expansion in a Plane Wave Basis

The conventional basis set for performing calculations of the electronic band structure of periodic materials, such as bulk semiconductors, is the plane wave basis.

$$\psi_i(\mathbf{r}) = \sum_\mathbf{G}^{E_{cut}} c_\mathbf{G}^i e^{i\mathbf{G}\cdot\mathbf{r}}. \qquad (10)$$

Within this basis, one can easily utilize Fast Fourier Transforms to convert the wave functions from a real space to reciprocal space representation. The matrix elements of the Hamiltonian in Eq. (5) in the basis of Eq. (10) can be written as:

$$\hat{H}_{\mathbf{G},\mathbf{G}'} = \frac{1}{2}\mathbf{G}^2\delta_{\mathbf{G},\mathbf{G}'} + V_{\text{local}}(\mathbf{G}-\mathbf{G}') + V_{\text{nonlocal}}(\mathbf{G},\mathbf{G}'). \qquad (11)$$

The conventional variational approach to solving for the eigenstates of Eq. (11) is to minimize the energy $\langle\psi_0|\hat{H}|\psi_0\rangle$, of the groundstate wave function, ψ_0, by varying its expansion coefficients, $c_{\mathbf{G},0}$. To find higher states, one needs to orthogonalize each ψ_i to all previously converged energy eigenstates below it. This orthogonalization process scales as the third power of the number of states and so only small systems with up to a few hundred electrons can be solved in this manner.

To enable us to study heterostructure systems containing 10^6 electrons, we instead "fold" the spectrum of eigenstates about a specified reference energy and hence solve[8,9] for the eigenstates of the equation

$$(\hat{H} - \epsilon_{\text{ref}})^2 \psi_i = (\epsilon - \epsilon_{\text{ref}})^2 \psi_i, \qquad (12)$$

where ϵ_{ref} is a chosen reference energy. By placing ϵ_{ref} within the band gap of the quantum dot system, and close to the valence band maximum (VBM) or conduction band minimum (CBM) one is then able to calculate the top few valence states or the bottom few conduction states respectively. As quantum confinement effects act to lower (raise) electron (hole) levels in the quantum dot compared to the bulk, one can ensure ϵ_{ref} falls within the band gap of the dot simply by placing it within the bulk band gap of the dot material. By applying the $\hat{H}$ operator twice, it is easy to see that the eigenstates of Eq. (12) are also eigenstates of the Hamiltonian in Eq. (5). Therefore, the process of folding the spectrum does not introduce any additional approximations compared to the conventional N^3 scaling algorithms. It simply removes the need to calculate all the low energy eigensolutions and then orthogonalize to each of these states. By removing the need for this costly orthogonalization, the Folded Spectrum algorithm is able to scale linearly with the number of electrons in the system. A version of this Folded Spectrum Code has been developed[10] for parallel supercomputers which linearly scales up to hundreds of processors and is able to handle systems containing up to 10^7 electrons. Typically, a calculation requires 20–50 plane waves per atom in the system and so these parallel calculations are effectively finding selective eigenstates of a matrix of order 10^8.

Expansion in a Linear Combination of Bloch Bands

For larger quantum dot heterostructure systems containing several million atoms, using a plane wave basis set to expand the single particle eigenstates becomes too computationally demanding, even when the folded spectrum method is employed. Instead, we choose to use a more physically intuitive basis set, namely a Linear Combination of Bulk Bands (LCBB)[11]:

$$\psi_i(\mathbf{r}) = \sum_s \sum_{n,k} c^{(i)}_{s,n,k} u_{s,n,k}(\mathbf{r}) e^{i \mathbf{k} \cdot \mathbf{r}}, \qquad (13)$$

where $u_{s,n,k}(\mathbf{r})$ is the cell periodic part of the bulk Bloch wave function for structure, s, at the nth band and the kth k-point. As these states form a physically more intuitive basis than traditional plane waves, the number of bands and k-points can be significantly reduced to keep only the physically important bands and k-points. This method was recently generalized to strained semiconductor heterostructure systems[12] and to include the spin-orbit interaction.[13] To study an $In_xGa_{1-x}As$ quantum dot embedded within a GaAs barrier, we typically use an LCBB basis derived from a set of structures, s. This allows us to directly include the effects of strain on the basis functions. The structures used are (i) unstrained, bulk InAs at zero pressure, (ii) unstrained, bulk GaAs at zero pressure, (iii) bulk InAs subjected to the strain value in the center of the InAs dot, and (iv) bulk InAs subjected to the strain value at the tip of the InAs dot. For an $In_xGa_{1-x}As$ quantum dot system, where we expect the electron and hole states to be derived from around the bulk Γ point, the wave vectors, $\{k\}$, include all allowed values within a given cutoff of the zone center. For calculations of electron states, we find we only need to include the band index, n, around the Γ_{1c} point. For the hole states we also include the three bands around the Γ_{15v} point. As the number of k-points and bands in the LCBB basis is increased, the eigenstates converge to those calculated using a converged basis, such as the above plane wave basis with a large cutoff, E_{cut}. We can therefore use a plane wave, folded spectrum calculation as a reference to judge when our LCBB basis set is sufficient. We find that to study a typical quantum dot system a basis set containing 10 000 bulk bands produces single particle energies that are converged with respect to basis size, to within 1 meV.

2.5. *Calculation of "two-body" interactions*

Using screened Hartree Fock theory, the energy associated with loading N electrons and M holes into a quantum dot can be expressed[14] as:

$$E_{MN} = \sum_i -\epsilon_{h_i} m_i + \sum_{i<j} (J_{ij}^{hh} - K_{ij}^{hh}) m_i m_j$$
$$+ \sum_i \epsilon_{e_i} n_i + \sum_{i<j} (J_{ij}^{ee} - K_{ij}^{ee}) n_i n_j - \sum_{ij} (J_{ij}^{eh} - K_{ij}^{eh}) n_i m_j, \quad (14)$$

where the electron and hole levels are denoted by $e_0, e_1, e_2, \ldots$, and $h_0, h_1, h_2, \ldots$ respectively. The n_i and m_i are the electron and hole occupation numbers respectively such that $\sum_i n_i = N$ and $\sum_i m_i = M$. The ϵ_i are the single-particle energies of the ith state, J_{ij} and K_{ij} are the direct and exchange Coulomb integrals between the ith and jth electronic states. For example, using Eq. (14), in the strong confinement regime where kinetic energy effects dominate over the effects of exchange and correlation, an exciton involving an electron excited from hole state i to electron state j can be expressed as:

$$E_{ij}^{\text{exciton}} = (\epsilon_{e_j} - \epsilon_{h_i}) - J_{ji}^{eh} + K_{ji}^{eh} \delta_{S,0}. \quad (15)$$

The direct and exchange Coulomb energies, are defined[15] as:

$$J_{ijkl} = \iint \frac{\psi_i^*(\mathbf{r}_1)\psi_j(\mathbf{r}_2)\psi_k^*(\mathbf{r}_1)\psi_l(\mathbf{r}_2)}{\bar{\epsilon}(\mathbf{r}_1 - \mathbf{r}_2)|\mathbf{r}_1 - \mathbf{r}_2|} d\mathbf{r}_1 d\mathbf{r}_2,$$
$$K_{ijkl} = \iint \frac{\psi_i^*(\mathbf{r}_1)\psi_j(\mathbf{r}_2)\psi_k^*(\mathbf{r}_2)\psi_l(\mathbf{r}_1)}{\bar{\epsilon}(\mathbf{r}_1 - \mathbf{r}_2)|\mathbf{r}_1 - \mathbf{r}_2|} d\mathbf{r}_1 d\mathbf{r}_2, \quad (16)$$

where $\bar{\epsilon}$ is connected to a phenomenological, screened dielectric function,[16] $\epsilon(\mathbf{r}, \mathbf{r}', R)$ by:

$$\frac{1}{\bar{\epsilon}(\mathbf{r}, \mathbf{r}'')|\mathbf{r} - \mathbf{r}''|} = \int d\mathbf{r}' \epsilon^{-1}(\mathbf{r}, \mathbf{r}'; R) \frac{1}{|\mathbf{r}'' - \mathbf{r}'|}, \qquad (17)$$

where R is the diameter of the quantum dot being studied.

2.6. *Many-body, correlation effects*

The above Hartree Fock based expression [Eq. (14)] for the total energy neglects the effects of correlation. These can be introduced by considering the effects of interactions between different configurations.[17] This is achieved by removing the restriction that the excited electron and remaining hole can only occupy the specific single particle levels, i and j. Instead, we describe the excitonic wave function, Ψ_{exciton}, in terms of an expansion of Slater determinants, $\Phi_{v,c}$, representing an orbital occupation for the electron in conduction state c and hole in valence state v,

$$\Psi_{\text{exciton}} = \sum_{v=1}^{N_v} \sum_{c=1}^{N_c} C_{v,c} \Phi_{v,c}, \qquad (18)$$

where N_v and N_c denote the number of valence and conduction states included in the expansion of the exciton wave functions. In this notation the valence states are numbered from 1 to N_v in order of decreasing energy starting from the valence band maximum, while the conduction states are numbered from 1 to N_c in order of increasing energy starting from the conduction band minimum. The Slater determinant, $\Phi_{v,c}$, is obtained from the ground state Slater determinant, Φ_0, by promoting an electron from the (occupied) single particle, valence state ψ_v with energy ϵ_v to the (unoccupied) conduction state, ψ_c of energy ϵ_c.

$$\begin{aligned}\Phi_0(\mathbf{r}_1, \ldots, \mathbf{r}_N) &= \mathcal{A}[\psi_1(\mathbf{r}_1), \ldots, \psi_v(\mathbf{r}_v), \ldots, [\psi_n(\mathbf{r}_n)], \\ \Phi_{v,c}(\mathbf{r}_1, \ldots, \mathbf{r}_N) &= \mathcal{A}[\psi_1(\mathbf{r}_1), \ldots, \psi_c(\mathbf{r}_v), \ldots, [\psi_n(\mathbf{r}_n)].\end{aligned} \qquad (19)$$

Within this basis set of Slater determinants, $\Phi_{v,c}$, the matrix elements of the many-body Hamiltonian, $\mathcal{H}$, are:

$$\mathcal{H}_{vc,v'c'} = \langle \Phi_{v,c} | \mathcal{H} | \Phi_{v',c'} \rangle = (\epsilon_c - \epsilon_v) \delta_{v,v'} \delta_{c,c'} - J_{vc,v'c'} + K_{vc,v'c'}, \qquad (20)$$

where J and K are the direct and exchange Coulomb matrix elements defined in Eq. (16). We can then obtain the excitonic energy levels and wave functions in the quantum dot by solving the eigenvalue equation

$$\sum_{v'=1}^{N_v} \sum_{c'=1}^{N_c} \mathcal{H}_{vc,v'c'} C_{v'c'} = E C_{v,c}. \qquad (21)$$

Having obtained the many-body excitonic energies, E, and wave functions, Ψ, we use Fermi's Golden Rule to obtain the near-edge absorption spectrum, defined as:

$$\sigma(\omega) \propto \frac{1}{V} \sum_{\alpha} |M^{(\alpha)}|^2 \delta(\hbar \omega - E^{(\alpha)}), \qquad (22)$$

where v is the nanocrystal volume and $M^{(\alpha)}$ is a particular dipole matrix element which can be expressed as a linear combination of the matrix elements between the single particle electron and hole levels,

$$m^{(\alpha)} = \sum_{vc} C_{vc}^{(\alpha)} \langle \psi_v | \mathbf{r} | \psi_c \rangle, \qquad (23)$$

where ψ_v, ψ_c are single particle valence and conduction states.

3. Recent Applications

One of the long standing problems that has held back the development of accurate models for the energy states in semiconductor quantum dot heterostructures is the need to accurately determine the size, shape and composition of quantum dot samples. This problem is further compounded by the fact that while it is possible to determine the shape of dots using atomic force microscopy (AFM) *before* they have been capped with a GaAs "barrier", it is believed that the capping process itself induces the diffusion of gallium into the dots and diffusion of indium from the dots into the surrounding matrix. Hence, AFM data for the size and shape of *uncapped* dots is of only limited use in evaluating the quality of theoretical models for *capped* dots. Therefore a major determining factor in our choice of which systems to apply the above EPM techniques to has been the continuing experimental progress in the characterization of self-assembled quantum dots. The earliest quantum dot samples were believed to contain pure InAs, pyramidal structures, with {101} facets (see Fig. 2) forming 45° angles between the facets and the base. Following this interpretation of the structure, early calculations were also performed assuming a pyramidal

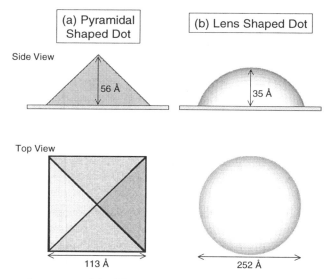

Fig. 2. The assumed geometry of the (a) pyramidal and (b) lens shaped dots described in Secs. 3.1 and 3.2.

geometry. In Sec. 3.1 we compare the results of our initial calculations for pyramidal shaped InAs quantum dots embedded within GaAs with those from other theoretical techniques. More recent characterization using cross sectional TEM and STM measurements of *capped* dots, has predicted that a more realistic dot geometry and composition is a lens shaped $In_xGa_{1-x}As$ quantum dot. In Sec. 3.2 we present results of calculations for the energy states, excitonic band gaps and Coulomb matrix elements in lens shaped $In_xGa_{1-x}As$ quantum dots. These calculations were used in combination with a range of measured properties to help to determine both the geometry and composition of these quantum dots.

The calculations described in Secs. 3.1 and 3.2 treat the interactions between electrons and holes at the level of first order perturbation theory, as in Eq. (15). For comparing with experimentally measured quantities such as the excitonic band gap and single particle level spacings, which have energies ranging from 0.1 to 1.0 eV this is certainly a sufficient level of approximation. In Sec. 3.3, we turn our attention to studying the energies of multi-exciton energy states in a quantum dot and calculate the energy associated with the decay from N to $N-1$ excitons. The *splittings* of these multi-exciton energy decays, e.g., the difference in energy between the decay of 3 excitons to 2 excitons and the energy of decay from 2 excitons to 1 exciton occupies a much smaller ~1 meV energy scale. Adopting the level of approximation described in Eq. (15) for these calculations would result in the omission of certain energy level splittings and actually calculate the wrong sign for some relative energies. For example, at the Hartree Fock level of approximation, the energy for the decay of a biexciton to a single exciton is *higher* than the energy of the decay of a single exciton. However, once the effects of correlation are included, the order is reversed and the biexciton decay energy is *lower*. Therefore in Sec. 3.3 we apply the more sophisticated CI formalism described in Sec. 2.6 to the problem of calculating multi-exciton energies in the same lens shaped dots discussed in Sec. 3.2.

3.1. *Pyramidal quantum dots: Single particle electron and hole states*

The EPM techniques described in Sec. 2 have been applied to the study of pyramidal self-assembled quantum dots in several publications.[16,18,19] The assumed pyramidal geometry is illustrated in Fig. 2(a). In Fig. 3 we show schematically the single particle energy levels that are typically calculated. The five lowest energy electron states are labeled in increasing energy from e_0 to e_4 and the three highest energy hole states are labeled in order of decreasing energy from h_0 to h_2. The single particle wave functions corresponding to each of these electron and hole states are illustrated in Fig. 4(a). Figure 4 shows only the square of the envelope function, $f(\mathbf{r})$, for each of the single particle states, i, defined as:

$$f^i(\mathbf{r}) = \sum_{\mathbf{k}} c_{n,\mathbf{k}}^i e^{i\mathbf{k}\cdot\mathbf{r}}, \qquad (24)$$

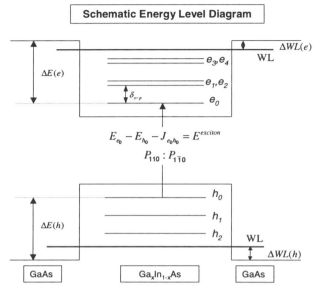

Fig. 3. A schematic representation of the single particle energy levels in a self assembled quantum dot. The five lowest energy electron states are labeled in increasing energy from e_0 to e_4 and the three highest energy hole states are labeled in order of decreasing energy from h_0 to h_2. The electron and hole binding energies are labeled $\Delta E(e,h)$ and the wetting layer energies with respect to the GaAs VBM and CBM are labeled $\Delta WL(h,e)$.

where the $c^i_{n,\mathbf{k}}$ are the coefficients of the ith single particle wave function, $\psi_i(\mathbf{r})$, when expanded in a basis of bulk Bloch orbitals representing a large number of bands, N, at the Brillouin zone center (Γ point),

$$\psi_i(\mathbf{r}) = \sum_{n=1}^{N} \sum_{\mathbf{k}} c^i_{n,\mathbf{k}} u_{n,\Gamma}(\mathbf{r}) e^{i\mathbf{k}\cdot\mathbf{r}}. \qquad (25)$$

The expansion coefficients, $c^i_{n,\mathbf{k}}$, are obtained by projecting each of the calculated single particle states onto this zone center basis,

$$c^i_{n,\mathbf{k}} = \langle \psi_i(\mathbf{r}) | u_{n,\Gamma}(\mathbf{r}) e^{i\mathbf{k}\cdot\mathbf{r}} \rangle. \qquad (26)$$

Plotting the envelope function squared, $f(\mathbf{r})^2$, instead of the wave function squared, $\psi_i(\mathbf{r})^2$, has the effect of averaging out the atomic scale oscillations in the wave function, purely for graphical purposes. In all our calculations, the atomistic structure of the EPM technique automatically includes both the envelope, $f(\mathbf{r})$, and atomistic, $u_{n,\Gamma}(\mathbf{r})$, contributions to the wave functions. A useful intuitive guide to interpreting the single particle electron states is to consider the eigenstates of the $\hat{L}_z$ operator.[20,21] The first six bound electron states corresponding to $l_z = 0, \pm 1$ and ± 2. The first state e_0, has $l_z = 0$ and is commonly described as s-like as it has no nodes. The e_1 and e_2 states have $l_z = \pm 1$, and are p-like with nodal planes (110)

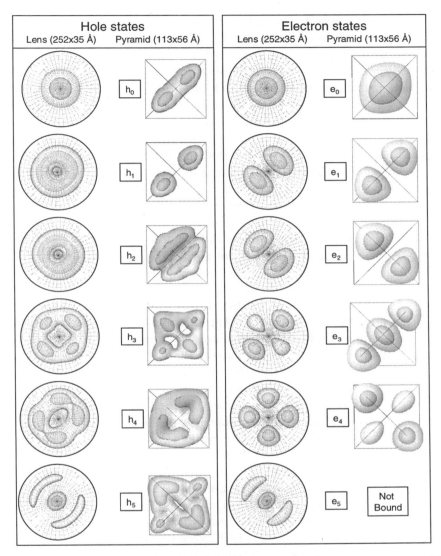

Fig. 4. Top view of the calculated electron and hole wave functions squared for pyramidal and lens shaped InAs quantum dots embedded in GaAs, with bases of 252 and 113 Å and heights of 25 and 56 Å. The light and dark gray isosurfaces represent 20 and 60% of the maximum charge density.

and $(\bar{1}10)$. The e_3, e_4 and e_5 states have $l_z = \pm 2$ and 0 respectively and are commonly described as $d_{x^2-y^2}$, d_{xy} and $2s$ respectively. Obviously, the square base of the pyramid reduces the symmetry of the dot from C_∞ to C_{4v}. In addition, the underlying zincblende atomistic structure, further reduces the symmetry to C_{2v}. Hence, the e_0 to e_5 states correspond to the a_1, b_1, b_2, a_1, a_2 and a_1 irreducible representations of the C_{2v} group, rather than eigenstates of $\hat{L}_z$. The alignment of the e_1 and e_2 p-states states along the [110] and [1$\bar{1}$0] directions results from the underlying zincblende lattice structure. Note, this simple analysis neglects the effects of the spin-orbit interaction which further reduces the symmetry from the C_{2v} group to a double group with the same single representation for all the states. In our calculations the spin-orbit interaction is included, but it produces no significant effects for the electron states. For the pyramidal dot studied in Fig. 4, the small size of the dot (113 Å base) and the relatively light effective mass of electrons in InAs (0.023 m_0) results in a large quantum confinement of the electron states, which in turn pushes these states up in energy so that only the five states e_0 to e_4 are bound in the pyramidal dot. Here, our definition of a bound electron (hole) state is that its energy is lower (higher) than the bulk GaAs CBM (VBM).

Figure 4 also shows calculated envelope functions squared for the hole states in the pyramidal shaped InAs/GaAs quantum dot. As there is a strong mixing between the original bulk Bloch states with Γ_{8v} and Γ_{7v} symmetry, the hole states cannot be interpreted as the solutions of a single band Hamiltonian. The larger effective mass for holes results in a reduced quantum confinement of the hole states and consequently many more bound hole states. Only the six bound hole states with the highest energy are shown in Fig. 4.

In Fig. 5 we show how the energy of the lowest four electron states and highest four hole states in a pyramidal InAs quantum dot depends on the size of the dot. It shows that as the size of the dot increases, the quantum confinement of the electrons and holes in the dot is reduced and therefore the electron levels decrease in energy and the hole levels increase in energy. Reducing the quantum confinement by increasing the size of the dots also acts to increase the number of bound states in the dot. For example, the e_3 state is bound for dots with a base size greater than 90 Å, but is higher than the wetting layer energy for dot with a base less than 90 Å and is therefore effectively unbound. It is interesting to compare these results with earlier calculations by (i) Grundman et al.[22] and Cusak et al.[23] who applied single band effective techniques to pyramidal InAs dots with base sizes ranging from 60 to 160 Å and found only a single bound state, and (ii) 8 band $\mathbf{k} \cdot \mathbf{p}$ calculations by Jiang et al.[24] and Pryor[25] who found three bound electron states over a similar range of sizes. We therefore conclude that to obtain even qualitative estimates for the quantum confinement energies in these systems requires a multi-band technique of at least eight bands.

3.2. Lens shaped dots: The effect of changing the shape and composition profile

In Sec. 3.1, we discussed the single particle electron and hole states in an idealized, pure InAs, pyramidal quantum dot. Recently, considerable evidence has emerged that, in fact, a more realistic geometry for self-assembled InAs/GaAs quantum dots is that of a lens shape dot[26–33] [see Fig. 2(b)]. In addition to modifying the predicted shape of self-assembled quantum dots, recent measurements[34–38] also indicate that

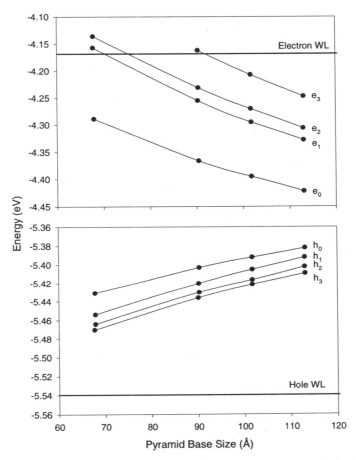

Fig. 5. The dependence of the energy of the lowest four electron states and highest four hole states on the base size of a pyramidal InAs quantum dot. The horizontal lines show the energy of the groundstate electron and hole energies in a 1 ML thick InAs WL.

the composition of these dots differs from the pure InAs that was originally assumed. In Table 3 we show the results of calculations for a pure InAs, lens shaped quantum dot, with a base of 252 Å and a height of 35 Å, embedded within GaAs [column (a)]. Table 3 also shows the experimentally measured splittings of the electron levels, the electron–electron and electron–hole Coulomb energies, the magnetic field dependence and the excitonic band gap measured in Refs. 28 and 32. The agreement between the measured energy level spacings, Coulomb energies and magnetic field response with our theoretical lens shaped model is generally good. Both the model and experiment find (i) a large spacing, δ_{sp}, (~50–60 meV) between the s-like e_0 state and the p-like e_1 state, (ii) a small spacing, δ_{pp}, (~3 meV) between the two p-like e_1 and e_2 states and (iii) a large spacing (~55 meV) between the p-like e_2 state and the d-like e_3 state.

Table 3. Calculated single particle electron and hole energy level spacings, electron and hole binding energies, $\Delta E(e, h)$, electron–electron and electron–hole Coulomb energies, excitonic band gap all in meV, exciton dipole moment and polarization anisotropy for lens shaped and pyramidal $Ga_xIn_{1-x}As$ quantum dots embedded within GaAs.

	Lens calculations						Pyramid calc.	Lens expt.
	(a)	(b)	(c)	(d)	(e)	(f)	(g)	
Geometry (Å)	252 × 35	275 × 35	252 × 25	252 × 35	252 × 35	275 × 35	200 × 100	Refs. 28, 32
% Ga at base, tip	0, 0	0, 0	0, 0	15, 15	30, 0	15, 15	0, 0	
$e_1 - e_0$	65	57	69	58	64	52	108	50
$e_3 - e_2$	68	61	67	60	63	57	64	48
$e_2 - e_1$	2	2	2	2	3	2	26	2
$e_2 - e_1$ (15T)	20	20	18	21	20	17		19
$e_4 - e_3$	4	3	4	4	3	1	23	
$h_0 - h_1$	8	12	16	13	14	11	15	
$h_1 - h_2$	7	6	5	5	6	5	20	
$h_2 - h_3$	6	10	14	13	14	9	1	
$\Delta E(e)$	271	258	251	209	192	204	171	
$\Delta E(h)$	193	186	174	199	203	201	198	
$J_{e_0 e_0}$	31	29	32	29	31	28	40	23
$J_{e_0 e_1}$	25	24	26	24	24	24	35	24
$J_{e_1 e_1}$	25	24	26	25	24	26	36	~18
$J_{h_0 h_0}$	30	27	39	32	28	30	31	
$J_{e_0 h_0}$	30	28	35	31	29	29	31	33.3
E_{gap}	1032	1016	1131	1080	1125	1083	1127	1098
d_{e_0, h_0} (Å)	0.16	−0.37	0.5	0.5	1.2	0.5	3.1	
$\lambda = P_{110} : P_{1\bar{1}0}$	1.03	1.01	1.04	1.05	1.08	1.08	1.20	

These electron level spacings are similar to those found for pyramidal quantum dots[18] [see Table 3, column (g)]. However, due to the lower pyramidal symmetry, the spacings of the two p-like and d-like states, δ_{pp}, δ_{dd}, are larger (26 and 23 meV). Both the model and experiment also find similar values for the Coulomb energies, $J(e_0 e_0)$ and $J(e_0 h_0)$ (~25 meV).

The calculated hole binding energy of $\Delta E(h) = 193$ meV is in good agreement with those of Berryman et al.[39] (~240 meV) and Itskevich et al.[40] (~250 meV). The calculated electron–electron and electron–hole Coulomb energies are in reasonable agreement with those extracted from Refs. 28 and 32. For the integrals $J^{ee}_{e_0 e_0}, J^{ee}_{e_0 e_1}, J^{ee}_{e_1 e_1}$ and $J^{eh}_{e_0 h_0}$ we calculate values of 31, 25, 25 and 37 respectively, compared to measured values of 23, 24, 18 and 33.3 meV. The calculated ratios of absorption intensities for light polarized along [110] and [1$\bar{1}$0] directions, defined as:

$$\lambda = \frac{P_{[110]}}{P_{[1\bar{1}0]}} = \frac{\langle \psi_{e_0}|r_{[110]}|\psi_{h_0}\rangle^2}{\langle \psi_{e_0}|r_{[1\bar{1}0]}|\psi_{h_0}\rangle^2}, \qquad (27)$$

are $\lambda = 1.03$ and 1.2 respectively for the $e_0 - h_0$ recombination in lens and pyramidal shaped, pure InAs dots.

In the lens shaped dot we find a difference in the average positions of the h_0 and e_0 states, d_{h_i,e_j}, of around 1 Å. This is smaller than the value we calculate for a pyramidal quantum dot, where we find the hole approximately 3.1 Å higher than the electron.

In summary, the assumed lens shaped geometry, with a pure InAs composition produces a good agreement with measured level splitting, Coulomb energies and magnetic field dependence. Detailed inspection of the remaining differences reveals that the calculations systematically *overestimate* the splittings between the single particle electron levels (δ_{sp}: 65 versus 50 meV, δ_{pd}: 68 versus 48 meV) and *underestimate* the excitonic band gap (1032 versus 1098 meV).

Pure InAs Dots: The Effects of Lens Shape and Size

Focusing on the lens shape only, we examine the effect of changing the height and base of the assumed geometry. Calculations were performed on similar lens shaped, pure InAs dots where (i) the base of the dot was increased from 252 to 275 Å, while keeping the height fixed at 35 Å [column (b)] and (ii) the height of the dot was decreased from 35 to 25 Å, while keeping the base fixed at 252 Å [column (c)]. These show that decreasing the height of the dot increases the quantum confinement and hence increases the splittings of the electron and hole levels (δ_{sp}: from 65 to 69 meV and δ_{h_0,h_1}: from 8 to 16 meV). Decreasing the height of the dot also acts to increase the excitonic band gap from 1032 to 1131 meV by pushing up the energy of the electron levels and pushing down the hole levels. Conversely, increasing the base of the dot decreases both the splittings of the single particle levels (δ_{sp}: from 66 to 61 meV) and the band gap (1032 to 1016 eV). These small changes in the geometry of the lens shaped dot have only a small effect on electronic properties that depend on the shape of the wave functions. The electron–electron and electron–hole

Coulomb energies remain relatively unchanged, the magnetic field induced splitting remain at 20 meV, the polarization anisotropy, λ, remains close to 1.0 and the excitonic dipole, d_{h_i,e_j}, remains negligible. In summary, reducing either the height or the base of the dot increases quantum confinement effects and hence increases energy spacings and band gaps, while not significantly affecting the shape of the wave functions.

Interdiffused In(Ga)As/GaAs Lens Shaped Dots

We next investigate the effect of changing the composition of the quantum dots, while keeping the geometry fixed. There have recently been several experiments[36,37,41] suggesting that a significant amount of Ga diffuses into the nominally pure InAs quantum dots during the growth process. We investigate two possible mechanisms for this Ga in-diffusion; (i) Ga diffuses into the dots during the growth process from all directions producing a dot with a uniform Ga composition $Ga_xIn_{1-x}As$, and (ii) Ga diffuses up from the substrate, as suggested in Ref. 41. To investigate the effects of these two methods of Ga in-diffusion on the electronic structure of the dots, we compare pure InAs dots embedded in GaAs with $Ga_xIn_{1-x}As$, random alloy dots embedded in GaAs, where the Ga composition, x, (i) is fixed at 0.15 [column (d)] and (ii) varies linearly from 0.3 at the base to 0 at the top of the dot [column (e)].

Table 3 shows that increasing the amount of Ga in the dots acts to decrease the electron level spacings (δ_{sp}: from 65 to 58 for $x = 0.15$). It also acts to increase the excitonic band gap from 1032 to 1080 and 1125 meV respectively. The electron binding energy, $\Delta E(e)$, is decreased by the in-diffusion of Ga (from 271 to 209 and 192 meV), while the hole binding energy, $\Delta E(h)$, is relatively unaffected. This significant decrease in the electron binding energy considerably improves the agreement with experiments on other dot geometries.[39,42]

As with changing the size of the dots, we find that Ga in-diffusion has only a small effects on properties that depend on the shape of the wave functions. The calculated electron–electron and electron–hole Coulomb energies are almost unchanged, while the average separation of the electron and hole, d_{h_i,e_j}, increases from 0.16 to 0.5 and 1.2 Å and the polarization ratio, λ, and magnetic field response are also unchanged.

Table 3 shows that the dominant contribution to the increase in the excitonic band gap and reduction in electron binding energy, results mostly from an increase in the energy of the *electron* levels as the Ga composition is increased. This can be understood by considering the electronic properties of the bulk $Ga_xIn_{1-x}As$ random alloy. The unstrained valence band offset between GaAs and InAs is ~50 meV,[43] while the conduction band offset in ~1100 meV and hence changing the Ga composition, x, has a large effect on the energy of the electron states and only a small effect on the hole states.

In summary, the effect of Ga in-diffusion is to reduce the spacing of the electron levels while significantly increasing their energy and hence increasing the band gap. We find that only the average Ga composition in the dots is important to their electronic properties. Whether this Ga is uniformly or linearly distributed throughout the dots has a negligible effect.

The effects of changing the *geometry* of the lens shaped, pure InAs dots on the single particle energy levels can be qualitatively understood from single band, effective mass arguments. These predict that decreasing any dimension of the dot, increases the quantum confinement and hence the energy level spacings and the single particle band gap will increase. Note that as the dominant quantum confinement in these systems arises from the vertical confinement of the electron and hole wave functions, changing the height has a stronger effect on the energy levels than changing the base. In this case decreasing the height by 10 Å has a much stronger effect on the energy spacings and on the band gap than increasing the base by 23 Å.

As increasing (decreasing) the dimensions of the dot acts to decrease (increase) both the level spacings and the gap, it is clear that changing the dot geometry alone will not significantly improve the agreement with experiment as this requires a simultaneous *decrease* in the energy level splittings and *increase* in the band gap. However, Ga in-diffusion into the dots acts to *increase* the band gap of the dot while decreasing the energy level spacings. Table 3 shows that adopting a geometry with a base of 275 Å and a height of 35 Å and a uniform Ga composition of $Ga_{0.15}In_{0.85}As$ produces the best fit to the measurements in Refs. 28 and 32.

In conclusion, our results strongly suggest that to obtain accurate agreement between theoretical models and experimental measurements for lens shaped quantum dots, one needs to adopt a model of the quantum dot that includes some Ga in-diffusion within the quantum dot. When 15% Ga in-diffusion is included, we obtain an excellent agreement between state of the art multi-band pseudopotential calculations and experiments for a wide range of electronic properties. We are able to fit/predict most observable properties to an accuracy of ±5 meV, which is sufficient to make predictions of both the geometry and composition of the dot samples.

3.3. *Multiple-exciton states in self-assembled quantum dots*

In the previous two sections we discussed the properties of single particle energy states within self assembled quantum dots. The nature of the empirical pseudopotentials included in the single particle Hamiltonian ensures that these states include exchange and correlation effects from the bulk materials to which they were fit. Therefore, bulk band gaps calculated from these single particle Hamiltonians will not suffer from the famous band gap problem that typically plagues density functional calculations of optical properties. However, there are two effects that are not included in these single particle states, (i) As the single particle Hamiltonian is not solved self-consistently, charge redistribution effects instigated by the excitation of an electron from the valence to conduction band are not included. In general, one expects all the single particle orbitals to change their shape when an electron in the system is excited from one state to another. This is a $1/N$ effect, i.e., the amount of relaxation of each orbital decreases with the total number of electrons in the system and is therefore usually small for such large dots. However, it may become significant when considering energy *differences* between similar multiexciton decays. (ii) When more than one electron is excited in the dot or the dot is charged with additional electrons, these electrons (and the resulting holes) will interact with each other via direct and exchange Coulomb interactions as well as a correlation interaction.

Both of these effects can be introduced using the configuration interaction (CI) technique described in Sec. 2.6. In this section we discuss the application of a CI expansion to examining the energy splittings between the decay of multiple excitons within self-assembled quantum dots. Recent, high-resolution single-dot spectroscopy[44–49] of InAs/GaAs self-assembled quantum dots has shown that at low power a single fundamental emission line is observed that is associated with the recombination of an electron from the e_0 level with a hole from the h_0 level as discussed in the previous sections. As the excitation intensity is increased, more excitons are loaded into the dots and new emission lines appear both to the red and to the blue of this fundamental emission line. Now, at higher excitation powers, there is an increase in the probability of observing "State filling" effects, in which the lowest electron and hole energy levels become completely filled and higher states start to become occupied. These higher excited electron–hole pairs states will recombine at energies, $e_i - h_j$. However, all these additional recombination energies are higher in energy than the fundamental emission lines are therefore cannot explain the red-shifted emission lines, nor the fact that the number of lines exceeds the number of allowed single-particle transitions. It is therefore been proposed[44–46] that multiple exciton decay is responsible for the rich detail observed in these single exciton spectra. Examples of this multi-exciton decay are schematically shown in Fig. 6, which shows the electrons and holes present when N excitons decay to $N-1$ excitons via the recombination of an electron in the e_0 state with a hole in the h_0 state. The vertical arrow shows the recombination taking place, and the blue and red spheres represent the electrons and holes. Only the three lowest (highest) electron (hole) levels are shown and each is assumed to have only spin (x2) degeneracy.

To isolate the physical factors contributing to multi-exciton effects we solve the problem in a series of steps. First, we neglect configuration interaction effects and treat only single-configurations. The $e_0 - h_0$ recombination energy in the presence of N_s electrons and holes is:

$$E_{e_0h_0}^{N\to N-1} = E_{e_0h_0}^{1/0} + \left[\sum_{e_s}^{N_s}(J_{e_0e_s} - J_{e_sh_0}) + \sum_{h_s}^{N_s}(J_{h_0h_s} - J_{e_0h_s})\right]$$

$$- \left[\sum_{e_s}^{N_s} K_{e_0e_s} + \sum_{h_s}^{N_s} K_{h_0h_s}\right], \qquad (28)$$

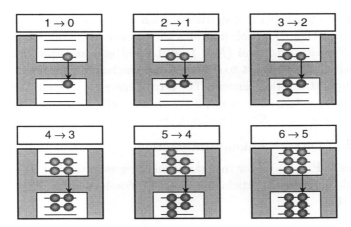

Fig. 6. Schematically representation of the radiative decay of N excitons to $N-1$ excitons via the recombination of an electron in the e_0 state with a hole in the h_0 state. The vertical arrow shows the recombination taking place, and the dark and light gray spheres represent the electrons and holes. Only the three lowest (highest) electron (hole) levels are shown and each is assumed to have only spin degeneracy.

where e_s and h_s are "spectator" electrons and holes, such that $\sum_{e_s}^{N_s} = \sum_{h_s}^{N_s} = N-1$. We see from Eq. (28) that the $e_0 - h_0$ recombination energy is shifted with respect to the fundamental exciton. This shift has two sources indicated by the two square bracketed terms in Eq. (28). First, there is a Coulomb shift given by the contents of the first bracket. This "Coulomb shift", $\delta^{\text{Coul}}_{N \to N-1}$ reflects the difference in the electron and hole wave functions and therefore vanishes if the electrons and holes, e_0 and h_0 have the same wave functions, i.e., if $J_{e_0 e_s} = J_{e_s h_0}$ and $J_{h_0 h_s} = J_{e_0 h_s}$. Second, there is an exchange shift, $\delta^{\text{exch}}_{N \to N-1}$, given by the second bracketed term in Eq. (28). This exchange shift is familiar from theories of band gap renormalization[50] where the existence of high carrier densities during high power photoexcitation act to reduce the band gap. In addition, since the exchange interaction depends on the spin orientation of the carriers, the exchange contribution can split the excitonic transitions.

Let us first consider the case where all the exchange integrals are zero, ($\delta^{\text{exch}} = 0$). In this ($\epsilon + J$) approximation we see only the effect of the Coulomb "chemical shift", δ^{Coul} due to the spectator electrons and holes. Second, we still keep only a single-configuration, but include exchange integrals. In this ($\epsilon + J + K$) approximation we will see the added effects of carrier–carrier exchange, δ^{exch}. Figure 7(a) shows our calculated energies associated with the fundamental recombination of an electron and hole, $e_0 \to h_0$ in the presence of 0 to 5 spectator excitons. These are denoted as the $1 \to 0$ to $6 \to 5$ transitions. The N excitons that form the initial state of each transition are assumed to occupy the groundstate configuration as predicted by the Aufbau principle, where the electron (hole) levels are sequencially filled in order of increasing (decreasing) from the band edges. The emission spectrum shown in each panel is constructed from a superposition of Gaussians. The position of each Gaussian peak indicates the calculated energy for the decay from an initial (N-exciton) to final ($N-1$ exciton) orbital and spin occupation. The width of each Gaussian is set to 0.1 meV and the heights are proportional to the calculated dipole transition element. The zero of energy is taken as the fundamental exciton, $\epsilon_{e_0} - \epsilon_{h_0} - J_{e_0 h_0}$. The multiplicity of each individual group of transitions in the $(\epsilon + J)$ and $(\epsilon + J + K)$ approximations are marked as solid and dashed lines.

The Effect of Direct Coulomb Interactions

The red lines in Fig. 7(a) show the transition energies in the $(\epsilon + J)$ approximation which include only single particle and direct Coulomb energies. Inspection of these lines shows that:

(i) All the observed Coulomb shifts are relatively small, ($\delta^{\text{Coul}} \sim 2$ meV) and result in a blue shift of the transition with respect to the fundamental transition. For example, at this level of approximation, the biexciton is "unbound" with respect to two single excitons, i.e., $E^{2 \to 1} > E^{1 \to 0}$.

(ii) As all the transitions involve the same single particle $e_0 \to h_0$ recombination, they all have the same oscillator strength.

(iii) All the transitions show only a single degenerate line as neither the initial or the final state exhibit any exchange splittings in this approximation.

The Effect of Exchange Interactions

The black lines in Fig. 7(a) show the same transition energies calculated in the $(\epsilon + J + K)$ approximation which includes single particle, direct and exchange Coulomb energies. Inspection of these lines shows that:

(i) The $1 \to 0$ and $2 \to 1$ transitions contain only one or zero spectator excitons and hence no electron–electron or hole–hole exchange takes place. We therefore classify these transitions as four-fold multiplets. Due to the lack of correlation the biexciton is still unbound with respect to two single excitons as in the $(\epsilon + J)$ approximation.

(ii) For *even* $\to$ *odd* there are four possible transitions. In each case, $2 \to 1$, $4 \to 3$ and $6 \to 5$ we observe only one four-fold multiplet. As the number of spectator excitons increases, there is a red shift of these transitions due to exchange interactions. The exchange energy shifts from Eq. (28) are $\delta^{\text{exch}}_{4 \to 3} = [K_{e_0 e_1} + K_{h_0 h_1}]$ and $\delta^{\text{exch}}_{6 \to 5} = [K_{e_0 e_2} + K_{e_0 e_1} + K_{h_0 h_2} + K_{h_0 h_1}]$.

(iii) For *odd* $\to$ *even* with $N \geq 3$ there are 64 possible transitions. In both cases: $3 \to 2$ and $5 \to 4$ we see six groups of transitions with multiplicities of $4 : 4 : 8 : 12 : 12 : 24$. The splitting between the six groups arises from electron–electron and hole–hole exchange splittings between the final states which contain two unpaired electrons and holes. The six groups of transitions in $3 \to 2$ and $5 \to 4$ span 23 and 22 meV. This energy span reflects the span of the 16 eigenvalues of the 16 × 16 matrix generated by the different spin occupations of the final states.

(iv) The exchange interaction alters the oscillator strength of the transitions so that they are not all identical as in the $(\epsilon + J)$ approximation.

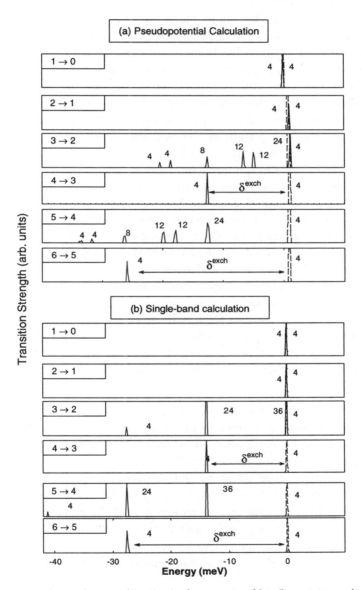

Fig. 7. The energy of $e_0 \to h_0$ recombination in the presence of 0 to 5 spectator excitons. Energies calculated in the $(\epsilon + J)$ (dashed) and $(\epsilon + J + K)$ (solid) approximations are shown. (a) Shows our pseudopotential calculations and (b) shows calculations with the assumptions from Refs. 44, 53 and 54. The multiplicity of each line is labeled.

Comparison with Alternative Techniques

Previous calculations of multi-excitons in quantum dots include the works of Hu,[51] Takagahara,[52] Barenco and Dupertuis,[53] Dekel et al.,[44,45] Landin et al.[46] and Hawrylak.[54] References 44, 45, 51–54 adopt single-band effective-mass models with either an infinite potential barrier[44,45,51–53] or a parabolic potential,[54] both of which artificially force the electron and hole wave functions to be identical. Therefore, in all these calculations the Coulomb shift, δ^{Coul}, is zero and the electron–hole exchange vanishes, $K_{e_i h_j} = 0$. To obtain realistic single particle energy spacings in Refs. 44, 45 and 53 an unrealistic cuboidal shape had to be assumed and the size of the dots was treated as adjustable parameters. By choosing different lengths for all three dimensions both the measured s–p and p–p splittings can be reproduced. The two-dimensional parabolic potential adopted in Ref. 54 can also be adjusted to reproduce the correct s–p splitting, but will always produce degenerate p states. In addition to approximating the single particle states, Refs. 44, 45, 51–54 neglect the effects of strain and the spin-orbit interaction.

To calculate the exchange and correlation contribution to the excitonic energies Ref. 51 uses path integral quantum Monte Carlo techniques which provide an exact (to within statistical error) determination of the exchange and correlation energy. However, quantum Monte Carlo methods are currently restricted to single band Hamiltonians and cannot therefore predict the Coulomb shift, δ^{Coul}, discussed above. Reference 54 adopts a limited basis CI to estimate correlation energies. References 44–46, 52 and 53 use only a single configuration approach which does not include correlation effects.

To assess the approximations used in Refs. 44, 45, 51, 53, and 54 we show in Fig. 7(b) a repeat of our calculations for the transition energies within the $(\epsilon + J)$ and $(\epsilon + J + K)$ approximations applying the assumptions adopted in Refs. 44, 45, 53, 54, namely $J_{e_i e_j} = J_{h_i h_j} = J_{e_i h_j}, K_{e_i e_j} = K_{h_i h_j}, K_{e_i h_j} = 0$ and $\Delta^{SO} = 0$. We observe that:

(i) Within the $(\epsilon + J)$ approximation, the chemical shift, δ^{Coul} is zero by definition so all the red lines lie on the zero of energy.
(ii) Within the $(\epsilon + J + K)$ approximation for *even* → *odd*, there are four possible transitions that are exactly degenerate as $K_{e_i h_j} = 0$.
(iii) For *odd* → *even*, there are 64 possible transitions, split into 4 : 24 : 36 multiplets, compared to the 4 : 4 : 8 : 12 : 12 : 24 multiplets obtained in the pseudopotential calculations. The reduction in the number of multiplets arises form the assumptions $\psi_{e_i} = \psi_{h_i}$ and $\Delta^{SO} = 0$. Other than these changes we find that many of the qualitative feature noted in the calculations of Dekel et al.[44,45] are retained in the pseudopotential description.

The Effect of CI Interactions

In Fig. 8 we contrast the transition energies from our pseudopotential calculations within the $(\epsilon + J + K)$ approximation (solid lines) with those from a CI calculation

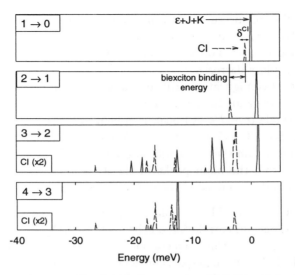

Fig. 8. Energy of $e_0 \to h_0$, recombinations in the presence of 0 to 3 spectator excitons. The solid and dashed lines show energies calculated in the $(\epsilon + J + K)$ and CI approximations.

(dashed lines) which includes single particle, direct and exchange Coulomb and "correlation" effects.

In the CI calculations we expand the many-body wave function in a basis of Slater determinants constructed from all possible orbital and spin occupations of the lowest 10 (including spin degeneracy) electron and highest 10 hole single particle levels. For example, the biexciton basis contains $^{10}C_2.^{10}C_2 = 2025$ Slater determinants, and the three-exciton basis contains $^{10}C_3.^{10}C_3 = 14400$ Slater determinants. This basis neglects the contributions for higher lying bound states and continuum states. To investigate the effects of adopting this limited CI basis[55] we have compared the results of diffusion quantum Monte Carlo calculations (DMC) and CI calculations for the exciton correlation energy in a model, single band system with equivalent size, band offsets and number of bound states. We find that our CI calculations retrieve approximately 50–60% of the 5–6 meV of correlation energy obtained in the DMC calculations and can therefore be used as a *lower bound* for the effects of correlation. Using these calculations we find that

(i) Correlation effects lower the energy of both the initial and final state of a transition. Our calculated correlation varies from 2–3 meV for a single exciton to 10 meV for multiple excitons. For all the transitions shown here, the correlation shift, δ^{CI}, for the initial state with N excitons is greater than that for the final $N - 1$ exciton state. Therefore, all the CI *transition* peaks are red shifted with respect to those from the $(\epsilon + J + K)$ approximation.

(ii) This (δ^{CI}) red shift is larger for the $2 \to 1$ transition than for the $1 \to 0$ transition and is able to overcome the Coulomb blue shift and "bind" the biexciton.

(iii) As the number of spectator excitons increases, the *difference* in the red shift for the initial and final states decreases, so that the red shift of the *transition* energy decreases. For $3 \to 2$ and $4 \to 3$ transitions the red shifts of the transitions rapidly decrease.

(iv) The mixing of configurations within the CI results in both additional peaks in the CI spectra and changes in the relative magnitude of the peaks, e.g., for $4 \to 3$ the single $(\epsilon + J + K)$ peak is split into two equally strong multiplets of peaks in the CI spectra. The additional peaks result from the mixing of excited states within the CI calculation. For example, the additional CI peaks blue shifted from the main peaks in the $4 \to 3$ spectra result from excited states configurations mixed into the $N = 3$ exciton.

In conclusion, we are able to isolate the effects of the direct and exchange Coulomb interactions and correlation on the energies of $N \to N-1$ excitonic transitions. We find that direct Coulomb energies introduce small blue shifts. Electron–electron and hole–hole exchange splittings which are responsible for the majority of the observed structure, introduce both red shifts and splittings. Correlation effects red shift all transitions and change the relative energies of transitions (e.g., bind the biexciton).

References

1. L.-W. Wang, A. Williamson, A. Zunger, H. Jiang, and J. Singh, *Appl. Phys. Lett.* **76**, 339 (2000).
2. P. Keating, *Phys. Rev.* **145**, 637 (1966).
3. C. Pryor, J. Kim, L.-W. Wang, A. Williamson, and A. Zunger, *J. Appl. Phys.* **83**, 2548 (1998).
4. R. Martin, *Phys. Rev. B* **1**, 4005 (1970).
5. L. Hedin, *J. Phys. C* **11**, R489 (1999).
6. A. Franceschetti, S.-H. Wei, and A. Zunger, *Phys. Rev. B* **50**, 17797 (1994).
7. Landolt and Borstein, *Numerical Data and Functional Relationships in Science and Technology, Volume 22, Subvolume a* (Springer-Verlag, Berlin, 1997).
8. L.-W. Wang and A. Zunger, *Semiconductor Nanoclusters* (Elsevier Science, Amsterdam, 1996).
9. L.-W. Wang and A. Zunger, *J. Chem. Phys.* **100**, 2394 (1994).
10. A. Canning, L. Wang, A. Williamson, and A. Zunger, *J. Comp. Phys.* **160**, 29 (2000).
11. L.-W. Wang, A. Franceschetti, and A. Zunger, *Phys. Rev. Lett.* **78**, 2819 (1997).
12. L.-W. Wang and A. Zunger, *Phys. Rev. B* **59**, 15806 (1999).
13. L.-W. Wang, A. Williamson, and A. Zunger, *Phys. Rev. B* (1999).
14. A. Franceschetti, A. Williamson, and A. Zunger, *J. Phys. Chem.* **60** (2000).
15. A. Franceschetti and A. Zunger, *Phys. Rev. Lett.* **78**, 915 (1997).
16. A. Williamson and A. Zunger, *Phys. Rev. B* **58**, 6724 (1998).
17. A. Franceschetti, H. Fu, L.-W. Wang, and A. Zunger, *Phys. Rev. B* **60**, 1819 (1999).
18. J. Kim, L.-W. Wang, and A. Zunger, *Phys. Rev. B* **57**, R9408 (1998).
19. L.-W. Wang, J. Kim, and A. Zunger, *Phys. Rev. B* **59**, 5678 (1999).
20. L. Jacak, P. Hawrylak, and A. Wojs, *Quantum Dots* (Springer, 1997).
21. A. Williamson, L.-W. Wang, and A. Zunger, *Phys. Rev. B* (2000).
22. M. Grundmann, O. Stier, and D. Bimberg, *Phys. Rev. B* **52**, 11969 (1995).

23. M. Cusak, P. Briddon, and M. Jaros, *Phys. Rev. B* **54**, 2300 (1996).
24. H. Jiang and J. Singh, *Appl. Phys. Lett.* **71**, 3239 (1997).
25. C. Pryor, *Phys. Rev. B* **57**, 7190 (1998).
26. H. Drexler, D. Leonard, W. Hansen, J. Kotthaus, and P. Petroff, *Phys. Rev. Lett.* **73**, 2252 (1994).
27. D. Medeiros-Ribeiro, G. Leonard, and P. Petroff, *Appl. Phys. Lett.* **66**, 1767 (1995).
28. M. Fricke, A. Lorke, J. Kotthaus, G. Medeiros-Ribeiro, and P. Petroff, *Eur. Phys. Lett.* **36**, 197 (1996).
29. B. Miller et al., *Phys. Rev. B* **56**, 6764 (1997).
30. R. Warburton et al., *Phys. Rev. Lett.* **79**, 5282 (1997).
31. K. Schmidt, G. Medeiros-Robeiro, J. Garcia, and P. Petroff, *Appl. Phys. Lett.* **70**, 1727 (1997).
32. R. Warburton et al., *Phys. Rev. B* **58**, 16221 (1998).
33. K. Schmidt, G. Medeiros-Robeiro, and P. Petroff, *Phys. Rev. B* **58**, 3597 (1998).
34. W. Yang, H. Lee, P. Sercel, and A. Norman, *SPIE Photonics* **1**, 3325 (1999).
35. W. Yang, H. Lee, J. Johnson, P. Sercel, and A. Norman, *Phys. Rev. B* (2000).
36. T. Metzger, I. Kegel, R. Paniago, and J. Peisl, *private communication* (1999).
37. J. Garcia et al., *Appl. Phys. Lett.* **71**, 2014 (1997).
38. M. Rubin et al., *Phys. Rev. Lett.* **77**, 5268 (1996).
39. K. Berryman, S. Lyon, and S. Mordechai, *J. Vac. Sci. Technol. B* **15**(4), 1045 (1997).
40. I. Itskevich et al., *Phys. Rev. B* **60**, R2185 (1999).
41. P. Fry et al., *Phys. Rev. Lett.* **84**, 733 (2000).
42. Y. Tang, D. Rich, I. Mukhametzhanov, P. Chen, and A. Hadhukar, *J. Appl. Phys.* **84**, 3342 (1998).
43. S.-H. Wei and A. Zunger, *Appl. Phys. Lett.* **72**, 2011 (1998).
44. E. Dekel et al., *Phys. Rev. Lett.* **80**, 4991 (1998).
45. E. Dekel, D. Gershoni, E. Ehrenfreund, J. Garcia, and P. Petroff, *Phys. Rev. B* **61**, 11009 (2000).
46. L. Landin et al., *Phys. Rev. B* **60**, 16640 (1999).
47. Y. Toda, O. Moriwake, M. Nishioda, and Y. Arakawa, *Phys. Rev. Lett.* **82**, 4114 (1999).
48. A. Zrenner, *J. Chem. Phys.* **112**, 7790 (2000).
49. A. Hartmann, Y. Ducommun, E. Kapon, U. Hohenester, and E. Molinari, *Phys. Rev. Lett.* **84**, 5648 (2000).
50. R. Ambigapathy et al., *Phys. Rev. Lett.* **78**, 3579 (1997).
51. Y. Hu et al., *Phys. Rev. Lett.* **18**, 1805 (1990).
52. T. Takagahara, *Phase Transitions* **68**, 281 (1999).
53. A. Barenco and M. Dupertuis, *Phys. Rev. B* **52**, 2766 (1995).
54. P. Hawrylak, *Phys. Rev. B* **60**, 5597 (1999).
55. J. Shumway, A. Franceschetti, and A. Zunger, *Phys. Rev. B* **63**, 155316 (2001).

SELF-ORGANIZED QUANTUM DOTS

A. R. WOLL, P. RUGHEIMER, AND M. G. LAGALLY
University of Wisconsin-Madison
Madison, WI 53706, U.S.A.

We review the concepts and principal experimental results pertaining to the self-assembly and self-ordering of quantum dots in semiconductor systems. We focus on the kinetics and thermodynamics of the formation and evolution of coherently strained 3D islands, and the effects of strain on nucleation, growth, and island shape. We also discuss ongoing research on methods to control the density, size, and size distributions of strained islands, both within a single strained layer and in quantum dot (QD) multilayers.

1 Introduction

The term "quantum dot" (QD) refers to a crystalline structure whose dimensions are small enough that its electronic states begin to resemble those of an atom or molecule rather than those of the bulk crystal. At this scale, the size of a crystal can be used to directly manipulate its electronic and optical properties. In principle, quantum dots with specific properties can be precisely tailored for specific applications, making them enormously useful in a wide variety of technologies. Yet, fabrication of QDs for technology is very difficult, requiring control on length scales on the order of 1-100 nm. Because of advances in techniques for materials fabrication and microscopy, research on quantum dots has greatly intensified in the last several years, involving academic and industrial research groups around the world.

Quantum dots within semiconductor materials have attracted particular attention because of the possibility of integrating dots with novel or superior properties into existing micro- and opto-electronic technologies. Currently, QD layers are being investigated for use in semiconductor lasers [1-4], storage devices [5,6], infrared detectors [7,8], and quantum computation [9]. Carrier confinement in such devices is achieved by embedding a smaller-bandgap material within a larger-bandgap material. Such confinement is routinely performed in *one* dimension by growing very thin, planar films using well-established technologies for thin-film growth, such as chemical vapor deposition (CVD) or molecular beam epitaxy (MBE). These methods allow practical control of layer thicknesses with accuracies on the Angstrom scale. Achieving confinement in all three dimensions requires lateral control on the same 1-100 nm length scale, and is much more difficult. For example, standard UV photolithography cannot yet produce such small features. While more exotic lithographic techniques, such as x-ray lithography or electron beam writing, do have adequate resolution to pattern QDs, they are comparatively expensive and not yet commonly used.

Because of the difficulty of QD fabrication using conventional techniques, a variety of alternatives have been investigated in which nano-sized particles form naturally, or "self-assemble", as a kinetic or thermodynamic response to their environment. For example, silicon layers that are buried at a precisely controlled depth within SiO_2, either by implantation or by layered film growth, coalesce into clusters with extremely narrow size distributions upon annealing [10]. This process is driven by the high interface energy between Si and SiO_2, and is shown schematically in Fig. 1(top). Porous silicon has been shown to consist of electronically isolated crystalline domains that show quantum size effects [11]. The small domain size is caused by the high rate of silicon deposition, resulting in a kinetically limited morphology. Another well-known example of self-assembly is the use of chemical manipulation to make highly monodispersed ensembles of metal [12–15] or semiconducting [16] particles. Such particles may then be chemically deposited onto a flat surface, forming well ordered, close-packed 2D arrays.

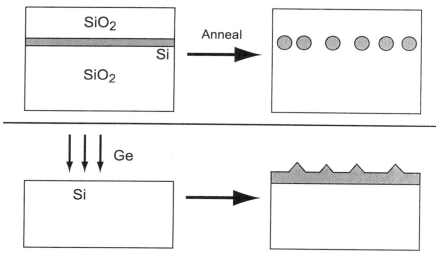

Fig. 1. (Top) Self-assembly of Si nanocrystals in SiO_2. First, a layer of Si is implanted into a SiO_2 layer at a well-defined depth. Because of the high interface energy, the Si layer forms isolated nanocrystals upon annealing. (Bottom) Strain-induced self-assembly of Ge nanocrystals on Si. Ge islands with well-defined facets form on top of a 2D film with a critical thickness of about 3 monolayers (ML).

An additional method for making self-assembled QDs takes advantage of the natural tendency of a strained, heteroepitaxial thin film to roughen, in some cases by forming 3D islands. This process is known as Stranski-Krastanov growth and is depicted in Fig. 1 (bottom). Although island nucleation on surfaces has been studied for many decades, it was only recently established that such 3D islands can remain epitaxially coherent with the substrate, i.e., free of dislocations [17], when the lattice-mismatch between the substrate and the film is not too large. This observation was extremely important, as dislocations within an island detrimentally affect its electronic properties and limit its usefulness as a quantum dot. Dislocation-free 3D islands have now been observed in a wide variety of group IV, II-VI, and III-V semiconductor systems. In each case, the film initially forms a smooth, 2D layer on the surface. After the film reaches a critical thickness, faceted, coherent 3D islands begin to nucleate. Such islands act as an alternative strain-relief mechanism to the formation of dislocations. While the island bases must be strained to match the substrate, the upper portion of the island can elastically relax towards its bulk lattice parameter.

Although strain provides a natural mechanism for the growth of dislocation-free 3D islands, many additional requirements must be met for such islands to be used as QDs. Because electronic properties are directly related to QD size and shape, it is important to understand how the physical properties of a QD are influenced by growth parameters, such as temperature, flux, and total coverage. In addition, the distribution of island sizes should be as narrow as possible. Finally, some potential applications require that dots have a specific arrangement with respect to one another. Because strain-induced elastic forces are long-range forces, dots interact strongly with each other, both within a single QD layer and between different layers separated by a spacer layer. "Self-organization" refers to anything that drives 3D islands towards greater regularity in position and/or size and shape than would ordinarily be expected in the absence of strain.

In this chapter, we review the details of semiconductor QD fabrication via Stranski-Krastanov growth. In section 2 we introduce strain-induced self-assembly in semiconductor heteroepitaxy, focusing on how strain influences the nucleation, size, shape, size distribution and thermodynamic stability of coherent islands. In section 3, we discuss a variety of approaches to influence, or engineer regularly sized QDs, and also to fabricate periodic QD arrays.

2 Quantum Dot Self-Assembly

Self-assembly of coherently strained, quantum-dot sized islands has been observed in a wide variety of semiconductor systems, including Ge/Si [17–20], CdSe/ZnSe [21,22], CdS/Zn(S)Se [23], PbSe/PbEuTe [24,25] (InGa)As/(AlGa)As [26,27], AlInAs/AlGaAs [28], GaN/AlN [29,30], InGaN/GaN [31], InP/InGaP [32–34], and GaSb/GaAs [35]. Each of these systems exhibits the same general morphological evolution, called Stranski-Krastanov growth, in which the deposited film is initially two-dimensional, but then begins to form clusters once a thickness of 1–3 monolayers is reached. This wetting-layer thickness is largely independent of growth rate and temperature [36], suggesting that the island formation reflects the equilibrium morphology of strained thin films. Any approach to quantum dot formation involving lattice-mismatched heteroepitaxy will therefore benefit from a full understanding of Stranski-Krastanov growth.

2.1 Equilibrium properties of coherent 3D islands

2.1.1 Equilibrium growth modes

When material of one type is deposited on another, the main factors affecting the equilibrium shape of the deposit are the surface free energies of the substrate γ_1 and the deposited material γ_2, and their interface energy γ_{12} [17, 37]. Considering these as the only terms in the total free energy leads to only two equilibrium growth modes. If $\gamma_1 > \gamma_{12} + \gamma_2$, then the equilibrium shape will be a flat film, the Frank-van der Merwe (FvdM) [38] growth mode. If, on the other hand, $\gamma_1 < \gamma_{12} + \gamma_2$, then the deposited material will form 3D islands. This growth mode is referred to as Volmer-Weber (VM) [39] growth.

In heteroepitaxial growth, the effect of strain, caused by lattice mismatch between the substrate and film, cannot be neglected. Lattice mismatch ϵ is defined as $(a_2 - a_1)/a_1$, where a_1 and a_2 are the lattice parameters parallel to the interface of the substrate and film, respectively. Generally, the effect of lattice mismatch is similar to that of interface energy. Very small values of ϵ tend to result in FvdM growth. When the mismatch is large, VW growth is favored. In this case, dislocations form between the island and substrate to accommodate the mismatch.

For intermediate values of ϵ, a third possibility exists. Suppose $\gamma_1 > \gamma_{12} + \gamma_2$ as is the case for FvdM growth. Then the deposited material will initially prefer to wet the surface. However, as the thickness t increases, an additional term, namely the strain energy of the film $\mu_2(t)$, must be added to the total free energy. When a critical thickness t_c is reached such that $\gamma_1 < \gamma_{12} + \gamma_2 + \mu_2(t_c)$, the equilibrium state will no longer be a smooth, coherent layer. In some systems, further deposition results in the formation of dislocations, while the morphology of the layer remains smooth. Alternatively, the extra strain energy can be relieved through the formation of 3D islands. Such systems are classified as following a third growth mode, called Stranski-Krastanov (SK) growth [40]. Prior to the mid-1980's, it was generally assumed that islands in SK growth relieved strain just like those in VW growth, by forming dislocations [17, 37, 41, 42].

In 1990, two groups [17, 43] reported TEM results on two different lattice-mismatched systems, Ge/Si(100) and InAs/GaAs, showing 3D islands in the epilayer that were coherent with the substrate. Eaglesham *et al.* [17] proposed that 3D islands, in at least some SK systems, relieve strain elastically. For Ge/Si(100), lattice planes in and near an island bend in order to accommodate the difference in lattice parameter between the substrate and film. The elastic stress on the island imparted by the substrate is mirrored by a stress imposed on the substrate by the island, resulting in a long-range strain field in the substrate. Such strain sharing is represented schematically in Fig. 2.

The existence of dislocation-free islands, and their ability to relieve strain by elastic deformation, has now been established in a variety of systems. Because of this, SK growth is now typically synonymous with the growth of a wetting layer

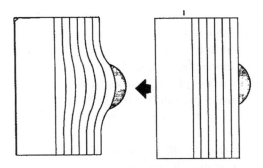

Fig. 2. Schematic model illustrating elastic strain relaxation of a lattice-mismatched, 3D island by local deformation of lattice planes in and near the island. From [17] with the kind permission of the authors.

followed by nucleation of elastically strained islands, rather than islands containing dislocations.[a]

Direct confirmation that islands may relieve strain elastically was obtained using *in situ* x-ray diffraction during Ge/Si(001) growth [44]. In these experiments, the in-plane lattice parameter was observed to shift towards the bulk Ge value at a coverage of 4 monolayers (ML), i.e., above the critical thickness for island formation but below the coverage at which dislocations appear [44]. Alternatively, one may measure the stress of a thin film, rather than the strain, by measuring the curvature of the thin film and substrate.[b] Because stress is the driving force for strain relaxation, this technique can be used to compare the effectiveness of different strain-relaxation mechanisms. In SiGe/Si, Floro *et al.* found that the film stress relaxed by as much as 20% compared to a planar film through the formation of pyramidal shaped islands("huts") [45, 46]. Furthermore, "dome" shaped, coherent SiGe islands, which form from large huts, are far more effective than huts at relieving stress, reducing it by over 60% compared to a flat, coherent film [46].

The difference in strain relaxation by domes and huts in Ge/Si(100) demonstrates a general relationship between an island's shape and its ability to relax strain. In general, island shape is determined by the competition between surface energy and strain relaxation in the island bulk. An island with steeper facets and a higher aspect ratio (height/base) relaxes more strain, but has more surface area, than one with a comparatively low aspect ratio [47, 48]. Germanium domes on Si(001), which are bounded by {113} and {102} facets [49], relax more stress than pyramids, bounded by relatively shallow {105} facets [18]. The change in shape from pyramid to dome as pyramids grow in size is discussed in section 2.2.3.

Islands with well-defined facets are also observed in most other SK systems.

---

[a] In most systems, coherent islands form dislocations when they become large. "SK growth" usually implies that they initially form as coherent islands.
[b] The curvature of a thin film on a substrate is proportional to the product of its stress and its thickness [45, 46].

Early RHEED observations of InAs growth on GaAs(100) [50] indicated that coherent, 3D InAs islands were bounded by steeper facets than those bounding Ge huts on Si. Recent, high-resolution STM images [51] confirm this interpretation, showing InAs islands bounded principally by four {137} facets. In PbSe/PbTe(111) [24, 25], the rock salt crystal structure of PbSe, along with the three-fold symmetry of the surface, lead to three-sided pyramids with {100} facets, with a contact angle of 54.7°. The aspect ratio of these islands is much higher than that in the (100) systems.

We emphasize that the classification of SK growth as an *equilibrium* growth mode is based on observations [36] showing that the critical thickness t_c is largely independent of kinetic parameters such as growth rate and temperature. However, there are also many examples of 2D-to-3D transitions in strained thin films that are kinetically driven [52]. For instance, Ag deposition on Pt(111) changes from 2D to 3D growth at 1 ML at 130 K and 6-9 ML at 300 K [52]. The 2D to 3D transition can be completely suppressed by growing at temperatures above about 450 K. The transition is caused by the presence of an additional diffusion barrier for interlayer transport, which can be overcome at sufficiently high growth temperature [52] or modified by the addition of a surfactant [53].

2.1.2 Coherent islands vs. dislocations

In a strained thin film, coherent 3D islands and dislocations are competing mechanisms of strain-relaxation. This competition may be described in terms of both kinetics and thermodynamics. The thermal activation barrier for the nucleation of islands scales as ϵ^{-4}, while that of dislocations scales as ϵ^{-1} [47]. Thus, for sufficiently high lattice-mismatch, coherent 3D islands will be kinetically favored over dislocations. The experimental observation of coherent, 3D islands serves as direct confirmation that they can be kinetically preferred. Whether such islands can also be favored thermodynamically is an entirely independent question, and is more difficult to address. It is convenient to break this question into two parts. First, can a single elastically strained island be more stable than a dislocated island of the same size? In other words, what is the lowest-energy configuration of a small, 3D island? This question is particularly important at the early stages of growth, when islands first begin to form and are necessarily small. Second, can a thin film with an ensemble of small, elastically strained islands represent the true equilibrium state of the system, compared to a film containing a few large islands in addition to dislocations?

The first of these questions has been theoretically analyzed [48, 54, 55] by comparing the relevant surface and interface energies: the free energy of the film surface, that of the dislocated interface, and the decrease in free energy from elastic relaxation in a coherent 3D island. A phase diagram resulting from this analysis is shown in Fig. 3. When the 3D island surface free energy is small relative to that of the dislocated interface, coherent 3D islands are favored. However, as these islands grow beyond a certain size, they become unstable to the formation of a dislocation [48]. Qualitatively, this size dependence is well matched by experiments. In both Ge/Si [18,56] and InAs/GaAs [48], 3D islands that grow too large eventually form dislocations.

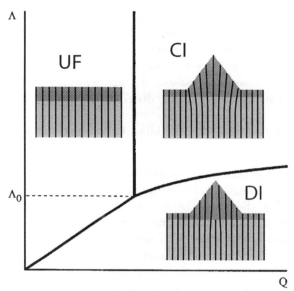

Fig. 3. Phase diagram showing the preferred morphology as a function of the amount of deposited material Q and the ratio $\Lambda = E_{interface}^{disl}/\Delta E_{surf}$ of the energy cost $E_{interface}^{disl}$ of a dislocated interface compared to the energy ΔE_{surf} due to the extra surface area imposed by the island. The labels UF, CI, and DI refer to "uniform film," "coherent island," and "dislocated island," respectively. From [54] with kind permission of the authors.

Such considerations imply that a small, coherent 3D island is in thermodynamic equilibrium so long as it is below a certain size. They do not address what happens when this constraint is lifted. For example, suppose further material is deposited onto a strained thin film after an island has already formed. Thermodynamically, is it preferable for these additional atoms to attach to the existing island, or instead to nucleate a new, small island?

If only surface free energies are important, the total free energy is minimized when there is only one large island. This is because spreading that material among many small islands increases the surface-to-volume ratio. Small 3D islands will thus have a tendency to rearrange into fewer, larger islands. Such spontaneous rearrangement is called coarsening, or ripening. In strained films, this coarsening takes on special significance. Suppose coarsening spontaneously occurs in an SK system and that dislocation formation is governed by the single-island phase diagram in Fig. 3. If allowed to grow, all coherent islands (CI) will eventually cross into the dislocated island (DI) region of phase space. Thus if coarsening is favored in a particular SK system, it is virtually assured that the system will eventually form dislocations.

Alternatively, it is possible that strain, an additional term in the free energy of the island, strongly influences the coarsening behavior. Models of SK growth that include the contribution of the island edge in the island free energy [57–59] predict that strain can counteract the driving force for coarsening, namely reduction of total surface area. An array of 3D coherent islands can, in principle, be energetically favored over a single, large island. If coarsening is not thermodynamically favored for a set of coherent, 3D islands, annealing such islands might narrow the island size distribution without increasing the average island size. One could then independently manipulate both the absolute size and size distribution of a set of islands, fine tuning their properties for quantum dot applications. In the next section we discuss the effect of strain on island size distributions in more detail.

2.1.3 Strain and the equilibrium island size distribution

Stranski-Krastanov growth is just one of many examples of strain-induced roughening in thin films. It is related to the well-known Asaro-Tiller-Grinfeld (ATG) instability [60,61], a long-length-scale, continuum model of strain-induced roughening. In the ATG approach, the energy of an entire thin film is evaluated as a function of the periodicity of regular corrugations on the surface. For certain combinations of materials parameters and kinetics, a particular wavelength corrugation will be favored, and eventually dominate the surface morphology. Such an evolution has been well documented experimentally, for example, in the growth of low-Ge-content $Si_{1-x}Ge_x$ alloys on vicinal Si(001) [62]. As for a single island, the stability of such a corrugation is determined by a competition between surface free energy and elastic strain-relaxation. Because of its continuum nature, the ATG approach is ill-suited to considering the thermodynamics of a small, crystalline island. For the rest of this section, we focus on the calculation of the strain energy of a single island, including the contributions of steps and facets.

Because elastic interactions are long-range, the strain distribution of an arbitrary set of coherent, 3D islands on a strained wetting layer is difficult to calculate analytically. However, elasticity theory can be used to estimate the nature and strength of different terms in the strain energy for a single island. By making use of such estimates, Daruka and Barabasi [57,58] calculated how a given amount of material will distribute itself on a substrate with a different lattice parameter. Atoms are allowed to occupy three "phases": the wetting layer, "small" islands, and large, "ripened" islands. The free energy per atom $E_{small}(x)$ of small islands depends on the island size x. That of ripened islands is the limit of E_{small} as $x \to \infty$. Equilibrium between these three phases is obtained by minimizing the free energy per atom of the entire system. The possibility of dislocation formation is not included in the calculation. Entropy is also neglected, so that the Helmholtz free energy is equivalent to the internal energy. The atomic free energy of the film plus island system is written [57]:

$$u = E_{wetting\,layer}(n_1) + n_2 \times E_{small}(x) + (H - n_1 - n_2) \times E_{ripened}, \qquad (1)$$

where H is the total amount of deposited material, and n_1, n_2, and $H - n_1 - n_2$

are the amount of material in each phase. For a given set of materials parameters, coverage H, and lattice mismatch ϵ, the equilibrium configuration is determined by minimizing u simultaneously with respect to n_1 and x. Minimization for different lattice mismatches and coverages results in phase diagrams giving the equilibrium occupancies n_1, n_2, and $H - n_1 - n_2$ [58]. A nonzero, equilibrium value for n_1 indicates the existence of islands that are stable against ripening.

It is worth describing the origin of the energy terms in Eq. 1 in more detail. $E_{wetting\ layer}$ depends on the strain energy of a smooth layer in addition to the bond energies between atoms in the deposited film and bond energies between atoms of the film and the substrate. The form of E_{small} for the energy per atom of a strained, coherent island was derived by Shchukin et al., and depends on the island size x:

$$E_{small} = \Phi + A\epsilon^2 + E_0\left(-\frac{2}{x^2}\ln(e^{1/2}x) + \frac{\alpha}{x} + \frac{\beta(n_2)}{x^{3/2}}\right). \qquad (2)$$

Eq. 2 was derived by dividing the total energy of a strained island by its volume, which is proportional to x^3. The first, constant term is the energy per atom of an unstrained film while the second is the (positive) strain relaxation energy per atom. The second term in parenthesis is proportional to the surface area of an island and reduces to the surface energy in the absence of strain. The other two terms in parentheses arise solely from strain. The first term is caused by the stress discontinuity at the island edge, while the last is a repulsive force between neighboring islands. $E_{ripened}$ is just the limit of E_{small} as $x \to \infty$.

Analysis of Eq. 1 for a particular set of materials parameters gives rise to a phase diagram describing the equilibrium morphology of the system as a function of layer thickness and misfit ϵ. One such phase diagram is shown in Fig. 4. The labels FM, VW, and SK correspond loosely to the principal growth modes described in section 2.1, but with several important differences. For example, the VW phase in Fig. 4 consists of coherently strained 3D islands, whereas the VW growth mode defined in section 2.1 is associated with 3D islands that contain dislocations. Also, the regions in which 3D islanding occurs, labeled SK, VW, and R, identify not just whether islands form, but whether islands have a thermodynamic preference to coarsen indefinitely, as discussed in section 2.1.2. For example, the SK_1 and SK_2 phases contain non-coarsening islands in coexistence with a smooth wetting layer. In the R_1-R_3 phases, at least some portion of small islands will exhibit a tendency to ripen.

The phase diagram in Fig. 4 excludes the possibility of dislocation formation. Therefore, the 'stable' SK phases predicted may in fact be metastable compared to a film with dislocations. However, we recall from section 2.1.2 that dislocations are energetically unfavorable for islands smaller than some critical value. The large, ripened islands within the R_1-R_3 phases of Fig. 4 should have a strong tendency to form dislocations. On the other hand, the rate of dislocation formation within 3D islands might be negligible for a thin film occupying an SK region of Fig. 4.

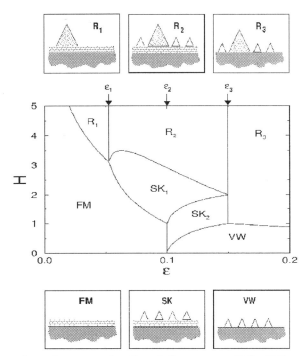

Fig. 4. Equilibrium phase diagram as function of the coverage H (in ML) and misfit $\epsilon = (a_1 - a_0)/a_0$ (unitless). The small panels on the top and the bottom illustrate the morphology of the surface in the six growth modes. The small empty islands indicate the presence of stable islands, while the large shaded one refers to ripened islands. ϵ_1, ϵ_2, and ϵ_3 delineate regions with qualitatively distinct morphological evolutions. Between ϵ_2 and ϵ_3, islands form first, followed by a wetting layer at the transition from VW to SK_2. Completion of this wetting layer is represented by the transition from SK_2 to SK_1. Figure reprinted with the kind permission of the authors [57].

2.2 Formation and evolution of coherent 3D islands

The size, shape, and density of quantum dots are all important when considering the fabrication of any real QD device. Because self-assembly occurs through island nucleation, a stochastic process, these parameters cannot be controlled directly. In order to optimize the properties of an ensemble of quantum dots, one must understand how island formation and evolution depend on growth parameters, such as temperature and growth rate. In this section we discuss the kinetics of coherent, 3D island formation and growth. To clarify the role of strain, we compare observations of SK growth with models that explicitly ignore strain and those that take strain into account. Where strain is important, we also discuss whether its influence on growth is primarily kinetic or thermodynamic. Experimentally, this distinction can be difficult to make.

2.2.1 Nucleation

In standard models of thin-film growth, nucleation of a 3D island from a supersaturation of adatoms on a surface represents a competition between the energy cost of creating additional surface area and the energy benefit of creating strong, chemical bonds in the island bulk. In a simple model of a 3D island with volume V, the free energy may be written as

$$E = aV^{2/3} - bV. \qquad (3)$$

The coefficients a and b represent the surface and bulk free energies of an atom in a 3D island relative to an adatom in the adatom sea. They determine the minimum island volume $V_c = (2a/3b)^3$ beyond which growth is energetically favored. The island energy given by Eq. 3 along with the chemical potential, dE/dV, are plotted in Fig. 5.

Strain in coherent, 3D islands modifies the free energy per bond in the island bulk. Because islands are partially relaxed, bonds among atoms within an island are stronger than that of atoms in the wetting layer, but weaker than it would be in unstrained, bulk material. In terms of Eq. 3, this modification is equivalent to a change in b.

More detailed analysis [47] corroborates the overall form of Eq. 3 for strained 3D islands. A critical assumption of this analysis is that islands have a constant shape. Experimental observations of the early stages of coherent-island formation show that this assumption may be inappropriate. For example, AFM measurements of island aspect ratios in $Si_{0.5}Ge_{0.5}$ [63] show the existence of small islands with contact angles as low as 4°. Larger islands all had 11.3° contact angles, consistent with {105} facets. Real-time low-energy electron microscopy (LEEM) measurements of low-concentration SiGe alloy film growth [64,65] show the formation of low-aspect-ratio mounds, whose contact angles gradually increase and eventually stabilize at 11.3°. In contrast with the simple nucleation model represented by Eq. 3, coherently

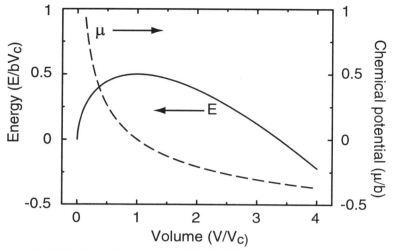

Fig. 5. Energy (solid line) and chemical potential (dashed line) of a single 3D island according to the model of Eq. 3. The critical 3D island volume is $V_c = (2a/3b)^3$, where a and b are positive constants.

strained, 3D islands in low-concentration $Si_{1-x}Ge_x$ alloys change shape as they form.

Even in the deposition of pure Ge on Si(001), the picture of nucleation of coherent, 3D islands on an atomically smooth wetting layer is too simple. Early STM observations [66, 67] show that, while the critical thickness for 3D island formation is 3 ML, the wetting layer begins to roughen at a thickness of 2 ML. At 2 ML deposition, 1 ML high 2D islands begin to appear on the surface. At 3 ML, a 1–2 ML height corrugation, or "checkerboard pattern" is observed on a lateral length scale distinct from that of steps caused by the substrate miscut. For deposition above 3 ML, recent studies [68] show that faceted, 3D islands are preceded by "prepyramidal" features, about 130 Å in diameter. These features have an aspect ratio of ≈0.04, less than half that of {105}-faceted huts.

Thus, formation of 3D islands of $Si_{1-x}Ge_x$ on Si(001), even for $x = 1$, is more complex that the simple model represented by Eq. 3. Evidently, non-faceted 3D islands form first, followed by a shape transition to faceted islands once they reach a certain size. However, the formation of such non-faceted initial islands should still take place via nucleation. *In situ* STM observations of pure Ge deposition on Si(100) [69] show direct evidence that islands initially form by a nucleation process. Small, 2D islands containing 120–270 atoms were observed to form quickly, but subsequently decay away. These islands were identified as "sub-critical" fluctuations, i.e., islands which form from a sudden, random accumulation of atoms, but without reaching the critical island size V_c.

2.2.2 Island coarsening: self-limited growth

The evolution of 3D islands in systems in which strain is not important is summarized by a few simple steps [70]. First, 3D islands nucleate from a supersaturation of adatoms created by an atomic flux onto the surface. These islands quickly deplete the adatoms in their vicinity, creating a "denuded zone", whose size is determined by adatom mobility. Once the denuded zones of neighboring islands overlap, very few additional islands nucleate. As long as atoms continue to land on the surface, existing islands will grow monotonically, capturing atoms that land either directly on them or within their denuded zones, until coalescence. The rate of capture of atoms by 3D islands increases with size. When the deposition flux is halted, the system attempts to lower its total surface area, and hence its free energy, via coarsening: atoms tend to leave small islands and diffuse to larger ones.

Experimental observations of island growth in many SK systems contrast dramatically with this simple evolution [71–75]. For example, under a constant deposition flux, the capture rate of atoms by 3D Ge islands on Si decreases down as they grow in size [71]. The growth rate of 3D InAs islands on InP(100) not only slows down, but changes sign with increasing coverage. The average island volume decreases by a factor of 5, accompanied by a dramatic increase in island number density, as coverage is increased between 1.5 and 2.0 ML.

The source of this phenomenon, often referred to as self-limited growth, has been rigorously investigated in recent years. Models to explain it fall into two categories, based either on thermodynamic considerations or solely on kinetics. At the heart of this debate is how much the total energy of a single island is modified by the presence of strain.

In section 2.1.3 we saw that a particular form for the free energy of a strained island, Eq. 2, led to the prediction of equilibrium phases in which multiple small islands are energetically preferred over a single, large island. The growth rate of 3D islands in such a system naturally slows down, or self-limits, as they grow above a certain size, at which point new islands nucleate. In general, Eq. 2 predicts a stable, non-coarsening SK phase when the free energy of the island edge is similar in magnitude to the free energy of the faceted island surface[c]. Alternatively, if the edge term is small, (and if the interaction term is neglected), then Eq. 2 reduces to the functional form of Eq. 3,[d] in which the only important terms are the bond energy and the surface free energy. Thermodynamically, such a system will try to reduce its surface area as in the simple picture of island evolution described above [70].

An island's growth rate can also slow down for kinetic reasons as it grows in size. In particular, strain-induced kinetic barriers can reduce adatom transport onto an island. The simplest kinetic model that leads to self-limiting growth depends only on general considerations of a strained, coherent island. The magnitude of the

---

[c]Both the edge and surface free energy terms depend nontrivially on strain. For a detailed discussion of the origin and role of these terms on island ordering, see Refs. [48, 59]

[d]Note that Eq. 3 is the free energy of an island with volume V, while Eq. 2 is the free energy per atom of an island whose volume is proportional to x^3

strain ϵ_s within this island is a maximum at the island edge. Because the chemical potential goes as ϵ_s^2 on a strained surface, it is also a maximum at an island edge, creating a kinetic barrier to diffusion. The strain, and hence this barrier, increases with increasing island size, leading to a decrease in the growth rate of 3D islands [76]. The decrease in growth rate allows smaller islands to catch up to larger ones.

An alternative kinetic model of self-limited growth is based on the observation that many coherent, 3D islands have well-defined facets. In this model, the kinetic barrier to island growth is due not to the cost of single adatom attachment but instead is associated with the nucleation of an additional layer on a faceted island boundary. The energy barrier for such a layer [18,71,77] increases with facet size. At sufficiently low temperatures, this barrier causes larger islands to grow more slowly than small ones, regardless of the thermodynamic preference of the island [18,71]. This model also explains the common observation, in Ge/Si(001), of elongated huts rather than square pyramids, despite experimental evidence that symmetric islands are energetically favored [78]. Suppose a perfectly symmetric, pyramidal island grows by one layer on one of its facets. This will cause the two adjacent facets to become trapezoids rather than triangles. Kinetically, further facet nucleation is disfavored on these trapezoidal facets compared to the facet that just grew. This kinetic instability tends to make pyramidal island become elongated [71].

Experimentally, determining whether self-limited growth is caused by kinetic or thermodynamic factors is difficult. Both factors can result in island growth rates that slow down with increasing size. However, self-limited growth in Ge/Si(001) appears most prominent at lower temperatures [71], suggesting that self-limited growth, in this case, is purely a kinetic effect.

In self-limited growth on InAs on GaAs(001), the distinction between kinetic and thermodynamic factors appears to be even more difficult. As in InAs/InP(001), the average island size appears to decrease with increasing coverage [73,74]. Figure 6 [73] shows histograms of the lateral and vertical island size distributions of InAs islands on GaAs at increasing coverages. The top two panels, corresponding to a total InAs coverage of ≈ 1.58 ML and an island density of $11 \mu m^{-2}$, show that islands have a relatively narrow size distribution centered on a lateral size of 180 Å and a height of 30 Å. For slightly higher coverage, corresponding to an island density of $24 \mu m^{-2}$, both the lateral and vertical island size distributions become very broad. With further coverage, the size distribution becomes narrow again. These first three rows of Fig. 6 may be understood as arising from self-limited growth. Islands initially nucleate at about the same time in film deposition, and are therefore all roughly the same size. As these islands proceed to grow, island nucleation continues, resulting in the broad distribution in the second row. Next, if the growth rate of large islands "self-limits", or decreases, small islands can catch up in size to large islands, causing the size distribution to become narrow once more.

The last three rows of Fig. 6 display an even more striking evolution. As the total density of dots increases, the lateral size of the dots actually decreases, while the height is roughly constant or decreases slightly. Thus the average island volume

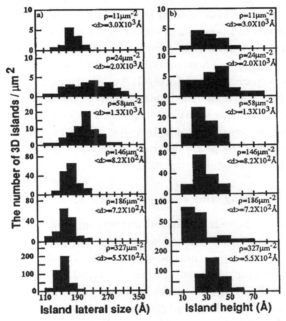

Fig. 6. InAs 3D island (a) lateral size and (b) height distribution for different total InAs 3D islands densities, ρ. Also indicated is the mean inter-island separation, $\langle d \rangle = \rho^{-1/2}$. The total InAs coverage ranges from $\approx$1.57 at the top $\approx$1.74 ML at the bottom. Reprinted from [73] with kind permission of the authors.

decreases with increasing coverage. Figure 6 also provides evidence that individual islands can decrease in size with increasing coverage. The number density of islands with lateral dimension above 230 Å decreases steadily beyond the third row of Fig. 6, suggesting that the growth rate of 3D islands not only decreases with increasing size, as in Ge/Si(001) growth, but reverses direction.

Further experimental evidence [73, 74] on InAs/GaAs(001) suggests that the decrease in island size with increasing coverage shown in Fig. 6 is associated with the increasing island density, and thus with island-island interactions. Empirically, the data suggest that an island that is surrounded by other, smaller islands has a tendency to donate atoms to these islands. This behavior stands in direct contrast with traditional phenomenology of island nucleation and growth described above. In such systems, the reduction of total surface area is the dominant driving force, so that mass transport between islands is always from small islands to large ones.

All explanations and models to account for the behavior shown in Fig. 6 are based on the role of strain. These models are both kinetic and thermodynamic in nature. Annealing experiments on InAs/GaAs islands after growth give additional experimental support for a thermodynamic model [48]. Annealing causes a narrowing in the InAs island size distribution without a decrease in island density [48]. These data support the possibility that InAs/GaAs(001) is an example of a "stable" SK phase in the sense of the phase diagram of Fig. 4: a system in which islands are thermodynamically stable against ripening.

2.2.3 Island coarsening and shape change

In quantum dot self-assembly, island shape commonly depends on size, both during growth and during 3D island coarsening. For Ge/Si(001), a very distinct transition from four-sided pyramids towards multifaceted "dome" shaped islands has been observed by many different groups [18,49,56,79]. This shape transition is frequently manifested as multiple peaks in the island size distribution [49]. In InP growth on GaInP(001) [80], the island size distribution has as many as three distinct peaks. Despite the early recognition that island shape transitions were associated with multi-modal size distributions, the complete physical description linking these phenomena has been a subject of some debate.

Ross et al. [56] explained the transition in Ge/Si(001) in terms of the dependence of 3D island free energy on the angle between the island and substrate. Figure 7 shows, according to their model, how the energy and chemical potential vary for two differently shaped islands as a function of size. The curves were generated by using Eq. 3 with different values of a/b, corresponding to different surface to volume ratios, or different shapes. When the energy of the more steeply faceted island becomes lower than the one with shallow facets, the equilibrium shape of the island changes. Although the energy difference between these two shapes is zero at the critical volume V_t, its derivative, the chemical potential, changes discontinuously. Such a transition is thus classified as first-order. Note that this model makes no explicit reference to strain.

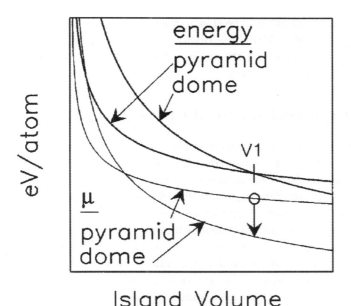

Fig. 7. Energy per atom and chemical potential as a function of island volume, according to Eq. 3, for dome-shaped and pyramid-shaped Ge islands on Si(001). The different shapes are modeled with different values of a/b [56]. Reprinted with the kind permission of the authors.

Medeiros-Ribeiro et al. [49, 78] proposed an alternative model for the hut-to-dome transition observed in Ge/Si(001), namely that the hut and dome islands each occur at a strain-induced local minimum in free energy with respect to island volume. This model does account for the existence of two distinct peaks in the island size distribution, but contrasts directly with the model of Ross et al., shown in Fig. 7, in which the free energy decreases monotonically. The model of Medeiros-Ribeiro is based on the theoretical analysis discussed in section 2.1.3 that illustrates how strain can modify the free energy of an island. The transition from hut to dome is viewed as a thermally activated process, involving the fast accumulation of adatoms from the surrounding surface.

The two competing models just described each explain the hut-to-dome transition in Ge/Si(100). Ross et al. obtained additional clarifying information by performing SiGe/Si(001) growth while observing the surface using LEEM [81]. They show that the transition from hut to dome takes place slowly, via a number of other island shapes, each of which is quite stable at the growth temperature. Furthermore, these shapes transformed reversibly to either huts or domes when the temperature was reduced, explaining why they were not observed in previous microscopy measurements conducted at room temperature. Theoretical support that the hut-to-dome transition is first-order in island volume was provided by Daruka et al. [82], who developed a 2D theoretical model of shape transitions in strained islands. The principal component of this model is the fact [47] that steeper facets allow greater elastic relaxation. The slow nature of the transition from hut to dome conflicts with the model of Medeiros-Ribeiro *et al.*, which predicts that the transition from hut to dome is a fast, thermally activated process involving the sudden capture of a large number of atoms.

3 Self-Organization of 3D island "Quantum Dots"

Most proposed electronic or opto-electronic devices involving quantum dots require that a large number of quantum dots be made with identical or nearly identical properties. In this section we focus on ways to manipulate and optimize an ensemble of 3D islands. We discuss self-organization and methods of control within a single layer of 3D islands, drawing on the ideas presented in section 2. We then focus on the growth of quantum dot multilayers: multiple layers of 3D islands, separated by spacer layers, consisting of the substrate material, which completely cover the islands so that the surface becomes almost as smooth as the original substrate. The lateral compositional inhomogeneity caused by the buried islands is transmitted to the surface through strain, so that nucleation of the second and subsequent quantum dot layers is no longer spatially random. Quantum dots prepared this way exhibit increasing degrees of order, size, shape, and position as more layers are added.

3.1 Single Layers

The characteristics of 3D, coherent islands that most influence their electronic and optical properties are size, size uniformity, and number density. Independent control of these parameters would allow quantum dot layers to be tailored for specific devices, but represents a significant challenge. In the following sections we describe work aimed at achieving such control.

3.1.1 Size and size distributions

The size distribution of coherent 3D islands can be quite narrow, but is very sensitive to growth parameters. For example, using AFM, Drucker *et al.* [83] compared the base lengths of pyramidal Ge islands on Si(001). The mean base length was ≈ 70 nm, while the full width at half maximum (FWHM) of the distribution of base lengths about the mean was 25%. Reducing the growth temperature from 550°C to 450°C caused the island base-length distribution to broaden to $\approx 40\%$. The narrow size distribution in Ge/Si has been attributed to kinetically self-limited growth, discussed in section 2.2.2.

Because of the shape transformation of Ge/Si 3D islands (described in section 2.2.3), it can be difficult to avoid a bimodal size distribution, in which both small "pyramids" as well as larger "domes" are present. Ross *et al.*, using *in situ* TEM, empirically determined a procedure to produce a monomodal size distribution of coherent domes [56] having a FWHM of only 15%. When domes first appear, their size distribution is relatively narrow. If deposition is stopped at this stage, and the Ge/Si film is annealed, all pyramidal islands will eventually disappear as their atoms diffuse to domes, whose chemical potential is lower (see Fig. 7). Once all pyramids disappear the size distribution is no longer bimodal, and the size distribution of domes is minimized. Further annealing results in coarsening of larger domes at the expense of slightly smaller ones, broadening the size distribution and, eventually, leading to dislocation formation. Thus the narrowest size distribution is obtained if the temperature is reduced just after all pyramids disappear.

For InAs/GaAs [73, 74] and PbSe/PbTe [25], typical size distribution widths are in the range of 7–15% of the mean island size for islands with mean bases sizes of ≈ 30 nm. The evolution of 3D coherent islands in these systems differs qualitatively from that of SiGe/Si in at least three important ways: First, the island size distribution in these systems is monomodal rather than bimodal. Also, the mean island size does not increase monotonically during growth [25, 73, 74]. Finally, the island size distribution in InAs/GaAs can become narrower during annealing [48]. These observations suggest that narrow island size distributions are more easily achieved in InAs/GaAs and PbSe/PbTe than in SiGe/Si.

As in the Ge/Si system, island size distributions in III-V systems are very sensitive to growth conditions. Fig. 6 in section 2.2.2 shows that differences of as little as 0.08 ML coverage result in a broadening of the size distribution from $\approx 10\%$ to 60% of the mean base size of InAs islands on GaAs. This sensitivity arises, evidently, from island-island interactions [73]. The island size distribution shown in Fig. 6 is broad in the early stages of growth, when islands are relatively sparse. As the island density increases, the mean island size and size distribution width both decrease. The precise relationship between island density and island size distribution is not yet understood.

3.1.2 Island density

The number density of coherent, 3D islands in SK systems varies qualitatively with temperature T and deposition flux F in the same way as nucleation in ordinary, VW systems [70]: the density goes as $(F/D)^\chi$, where D is the diffusion constant, which depends exponentially on T, and χ is positive and depends on the details of the system. Such a dependence has been observed both for SiGe/Si(001) [84] and for InAs/GaAs [85]. Thus density increases with increasing flux and decreases with increasing temperature. In SiGe/Si(001), the number density of coherent, 3D islands has been manipulated from about 10^8–10^{12} islands/cm^2 [19, 84, 86, 87]. The total 3D island density also depends sensitively on the total coverage within a certain coverage range. For InAs/GaAs, the appearance of 3D coherent islands is very sudden. At a growth rate of ≈ 0.01 ML/s, the 3D island density changes from zero at 1.5 ML to $\approx 10^{10}/cm^2$ at 1.7 ML [73, 74].

3.1.3 Lateral position

Many potential applications for QDs require precise control not only over their shape, size, and size distribution, but also over their lateral position. Recently, an increasing effort has been devoted to finding new methods to achieve such control over self-assembled, 3D coherent islands. Most of these entail the use of the substrate as a template for island nucleation, through lithography defined surface morphology [20, 88–93], surface steps or step-bunches [92, 94–97], or the intentional formation of dislocations [98].

Kamins et al. [89, 91] deposited Ge islands on raised Si strips on a (001) wafer. Valleys were etched through a SiO_2 layer down to the silicon and parallel to the $\langle 100 \rangle$ direction. Using CVD, Si was then selectively deposited on regions of exposed Si, resulting in raised Si strips bounded by {011} side walls. For strips varying in width from 450 nm to 1.7 μm, islands nucleated preferentially at the edges, forming a linear array, and had square bases with a mean side length of 86 nm and a mean height of 15 nm.

Jin et al. also investigated the arrangement of Ge islands on lithographically patterned sub-micron mesas [20, 88]. Figure 8 shows a variety of patterned mesas on Si(001) and the arrangement of 3D Ge islands grown on top of them. Long, raised stripes result in extremely uniform islands growing in long, ordered arrays, similar to the results of Kamins et al.. Deposition of 9 ML of Ge on raised, square mesas results in the formation of islands at each corner. Further deposition results in the formation of a fifth, smaller island in the center. These results demonstrate the feasibility of using lithography to control the position at which Ge islands nucleate on a Si substrate.

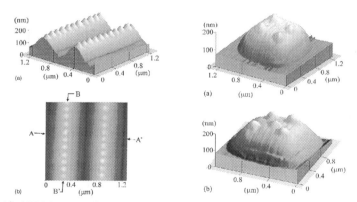

Fig. 8. (Left) AFM images of self-organized 3D Ge dome-shaped islands on a $\langle 110 \rangle$-oriented Si stripe mesa with a window width of 0.6 μm. (Right) Ge domes on square Si mesas with base lines parallel to the $\langle 110 \rangle$ direction. The Ge thickness is (a) 9 ML and (b) 10 ML. The average island base size is 140 nm. Reprinted from [20] with the kind permission of the authors.

Recently, Rugheimer et al. [93] investigated Ge deposition on raised Si mesas with dimensions of 3 to 20 μm on both bulk Si(001) and silicon-on-insulator (SOI) substrates. These mesas were fabricated by a single patterning step followed by a reactive ion etch, and are thus bounded by nearly vertical side walls, rather than low-index facets. At a deposition temperature of 700°C, the distribution of coherent, pyramidal islands during early stages of deposition appears to be unaffected by the mesa edge. Beyond a certain size, however, islands at the edges dominate the coarsening process. At this temperature, intermixing also occurs: large islands at the edge deplete not only Ge from nearby, smaller islands, but also Si from the surrounding substrate. For mesas fabricated on SOI with a ≈10 nm thick Si template layer, each large edge islands absorbs all the Si from the substrate surrounding it, creating an isolated ring of SiGe islands at the mesa edge [93]. Once these islands have locally depleted the Si template layer, the process repeats itself as a new ring of isolated SiGe islands is formed inside the first one. Figure 9 shows an AFM image of a mesa with three such "island rings". Cross-sectional TEM verifies that the 10 nm silicon layer between these 3D islands is completely removed down to the oxide, but remains intact underneath the 3D islands. TEM results also indicate that these large islands contain no dislocations.

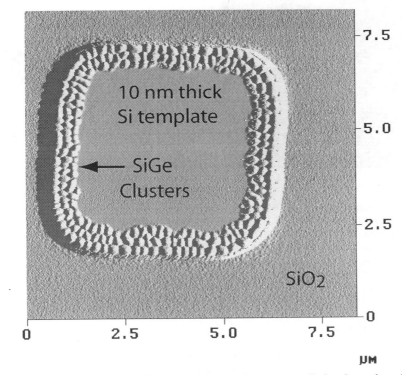

Fig. 9. AFM image of a raised 5 μm SOI mesa onto which 1.6 nm Ge has been deposited using MBE at a growth temperature of 700°C. The Si template layer is 10nm thick. The region around the mesa is oxide. The image is shown in differential mode.

The use of patterning to manipulate the lateral position of 3D, coherent islands has also been applied to III-V systems [92, 99]. Ploog et al. have demonstrated the controlled formation of quantum wires, quantum dots, and coupled wire-dot arrays of GaAs bounded by AlGaAs. Such control was achieved by combining lithographic techniques with methods to achieve step bunching on [311]A GaAs substrates.

Dislocations, like lithographically defined patterns, can act as preferred sites for island nucleation. Teichert et al [98] exploited this phenomenon to create a highly ordered square array of pyramidal, {105}-faceted SiGe islands. Starting with a clean Si(100) surface, they grew 30 nm of $Si_{0.7}Ge_{0.3}$ at 150°C in the presence of 1 keV Si^+ ions, followed by conventional growth of 50 nm of additional alloy at 550°C. The low temperature and ion bombardment during the first growth step results in the formation of a dense array of dislocations, which act as preferred sites for island nucleation. Figure 10 shows AFM images of the resulting island array. The ordering is most pronounced in regions of highest dislocation density, such as the 1×1 μm region enlarged at the right. It is possible that the size of such well-ordered regions might be significantly increased by further augmenting the dislocation density.

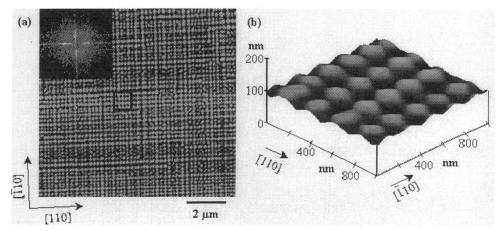

Fig. 10. Morphology of a 80-nm $Si_{0.7}Ge_{0.3}$ film, of which the first 30 nm was grown at 150°C under simultaneous 1 keV Si^+ ion bombardment. The remaining 50 nm was grown without ion bombardment at 550°C. (a) 10×10 μm image, gray-scale: 15 nm, inset: corresponding power spectrum ranging from -25.6 to 25.6 μm^{-1}. (b) Three-dimensional 1×1 μm image obtained with an electron-beam deposited tip of the area framed in (a). Reprinted from [98] with the kind permission of the authors.

3.2 Multilayers

Multilayers of coherent 3D islands not only offer opportunities to realize three-dimensional arrays of QDs, but also provide an additional tool to explore the atomistic mechanisms of self-assembly through strain management. These multilayers are fabricated by alternating heteroepitaxial layers with spacer layers of the substrate material. The spacer layer fills in the region between islands before covering them, so that the resulting surface is as smooth as the initial substrate. The lateral inhomogeneity due to buried islands is transmitted to the surface through strain, so that nucleation of 3D islands in the second and subsequent layers is no longer spatially random. By controlling the relative thicknesses of the two layers, the com-

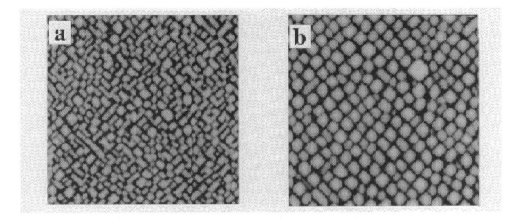

Fig. 11: AFM images of $Si_{0.25}Ge_{0.75}$/Si(001) superlattices. The horizontal direction corresponds to [110]. (a) 0.8×0.8 μm image of a single SiGe layer. Gray scale range is 5 nm. (b) 1.25×1.25 μm image of the 20th SiGe layer. Gray scale range is 10 nm. Reprinted from Ref. [100] with the kind permission of the authors.

position of the heteroepitaxial layer, the temperature at various stages of growth, and the deposition rate, 3D island sizes, shapes, size distributions, and spatial distributions can be modified and to some extent controlled. In all cases, 3D islands in multilayers are coherent with the substrate.

3.2.1 Basic concepts of multilayer ordering

The AFM images shown in Fig. 11 compare the arrangement of $Si_{0.25}Ge_{0.75}$ QD clusters in a single-layer film (a) and in the 20th SiGe layer of a Si-SiGe multilayer (b) grown on Si(001). All SiGe layers have a mean thickness of 2.5 nm, and are separated by spacer layers consisting of 10 nm Si. Analysis of both images reveals that the islands are bounded by {105} facets. The island size is larger in the multilayer film, with a concomitant decrease in island number density [100, 101].

The increased size of coherently strained 3D islands with increasing layer number is accompanied by an increased regularity, both in size distribution and position. The power spectrum of the images [101] shows a fourfold pattern in each case, arising from the short-range order of close-packed rectangular islands. However, the pattern from the multilayer surface is considerably sharper, and exhibits 2nd-order peaks, indicating a much more regular array. Quantitatively, the width of the island-size distribution, for these close-packed islands, may be estimated from the width of the first peak in the power spectrum along a {100} direction. It changes from $\sim 1.1 \langle L \rangle$ to $\sim 0.3 \langle L \rangle$ between Figs. 11a and 11b, where $\langle L \rangle$ is the average lateral size of the islands. It is also clear from the images that the island shape becomes more square for the multilayer film. The average aspect ratio (base length to width) of the islands changes from approximately 1.50 to 1.15 ± 0.05. Thus the

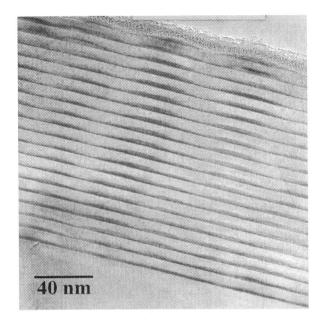

Fig. 12: (From Ref. [102]) Representative [110]-cross-section bright-field TEM micrograph of an uncapped 20 bilayer $Si_{0.25}Ge_{0.75}$/Si multilayer, showing the vertical ordering of the SiGe islands.

increased regularity evident in Fig. 11 may be characterized by a narrowing in the distribution of island sizes and spacings, and by an evolution from rectangular to square island shapes.

Cross-sectional TEM measurements elucidate the ordering mechanism operative in Fig. 11. Figure 12 shows a bright-field image [102] of a sample grown in a manner similar to that of Fig. 11b. Dark regions of the image correspond to SiGe alloy, while lighter regions correspond to the Si substrate and Si spacer layers. Islands appear to be stacked on top of one another, or vertically correlated. This phenomenon is a nearly ubiquitous feature of multilayer QD films [27, 30, 34, 100–107].

In order for lateral ordering in island positions to occur, the vertical correlation evident in Fig. 12 cannot be absolute. If it were, subsequent layers in a multilayer film could only replicate the stochastic nucleation pattern of the first layer. The middle column of islands in Fig. 12 shows evidence for a gradual change in the arrangement of islands as the number of layers is increased. In particular, small islands move towards one another [102]. Once the pair of islands becomes close enough, only one island of larger size appears in the following layer. Subsequently, this island replicates itself without significant further changes in size or position [102].

Tersoff *et al.* proposed a simple model to explain the ordering observed in Fig. 11 [100]. Islands are treated as spherical inclusions in a continuum elastic

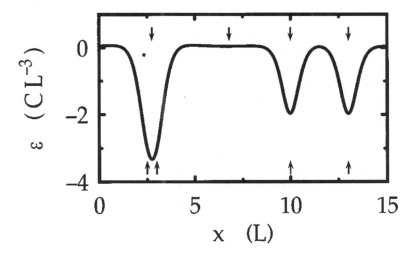

Fig. 13: Surface strain ϵ vs. lateral displacement x (in units of spacer layer thickness L), for four islands buried at depth L according to the calculation in Ref. [100]. Arrows at bottom indicate lateral positions of buried islands. Arrows at top indicate minima in ε, i.e., favored positions for subsequent nucleation. Reprinted from Ref. [100] with the kind permission of the authors.

medium of the substrate material buried at depth L below a film surface. The strain field at the surface is then calculated, and each minimum in the strain field with respect to the lattice parameter of the QD material is chosen for nucleation of islands in the current layer. In SiGe/Si multilayers, these regions will be under tensile strain relative to pure Si, such that the lattice parameter is closer to that of the SiGe alloy. The energy cost of nucleating a coherently strained island at these strain minima is smaller than in regions between islands, where the Si spacer layer is closer to its bulk equilibrium value [100].

Figure 13 shows the results of such a calculation: the strain field at the surface as a function of lateral position above a buried layer of islands. The two right-hand islands are far enough apart that they each give rise to a minimum in the strain field at the top, exposed surface. Islands that are very close together, such as the two left-most islands in Fig. 13, give rise to a single minimum, centered between the two islands, causing a single island to form above the island pair. The distance over which strain fields interact is determined primarily by the spacer-layer thickness [100]. Thus if this thickness is kept constant, island sizes must eventually reach a "stable" size beyond which they will no longer merge with one another.

This model can be used to explain almost all of the phenomena shown in Fig. 12. The approach of two islands toward each other, island mergers, and the stabilization of island sizes after such mergers are all expected from interactions of strain fields of buried islands. Simulations, based on the calculated strain fields and deterministic rules for nucleation, give similar results to Figs. 11-12 [100].

Using TEM, Kienzle et al. [105] experimentally determined the degree of vertical correlation as a function of spacer layer thickness in the Ge/Si system. They show that spacer layer thicknesses between 12.5 and 25 nm result in the largest percentage of vertically correlated islands. In addition, they observed a "critical" spacer layer thickness of 70nm, beyond which islands have a probability of less than 1/2 of being vertically aligned with buried dots. Similar results were obtained for InAs/GaAs by Xie et al. [27]. For this system, the probability for islands to be vertically correlated with islands in other layers begins to decrease at a spacer layer thickness of approximately 50 nm and falls to 1/2 at about 90 nm.

Vertical correlation can manifest itself in more complex ways than that shown in Fig. 12. For example, anticorrelation, in which islands in adjacent layers are laterally shifted with respect to one another, has also been observed [108]. In PbSe/Pb$_{1-x}$Eu$_x$Te multilayers [24], 3D islands in adjacent layers are positioned according to FCC stacking rules. These effects are qualitatively understood in terms of the model described above: preferred nucleation sites are determined by the strain field from buried islands. Because of the anisotropic elastic properties of real crystals, preferred nucleation sites are not necessarily located above buried islands [108].

3.2.2 Ordering in sparse arrays

The experiments and theory discussed above primarily involve multilayers with densely packed islands. For example, analysis of Fig. 12 and similar TEM data suggests that merging of islands is the dominant effect in the evolution of multilayer quantum dot films. These mergers appear to cause the increased uniformity in island size: the distribution of "daughter" islands (these are islands closer to the surface which lie above a pair of deeper islands) is noticeably narrower than that of the "parent" islands. However, multilayers of sparse arrays of QDs (i.e., having a low number density), in which island mergers are rare, can also exhibit a narrowing size distribution with increasing layer number. In these systems, the island density is approximately constant [27,34], so that the narrowing island size distribution cannot be explained by island mergers. In addition, because buried islands in the model represented by Fig. 13 are considered spherical, no explanation of the evolution from rectangular to square islands shown in Fig. 11 can be expected.

In order to explain these aspects of multilayer QD ordering, Liu et al. [109] performed deterministic simulations, resembling those in Ref. [100], but based on strain fields from more realistic shapes for the buried islands. One of the principal analytical results of this calculation is the dependence of the extent, x_0, of the tensile region of the strain field at the spacer layer surface on the width, W, of the buried island. In particular, x_0 is always larger than W and is proportional to $W^{1/2}$.

In their simulations, Liu et al. propose that the size of an island that nucleates in a region of tensile strain is fully determined by the size of this strained region. In particular, they fix the ratio of new island size, B_i, to tensile region, x_0, at some

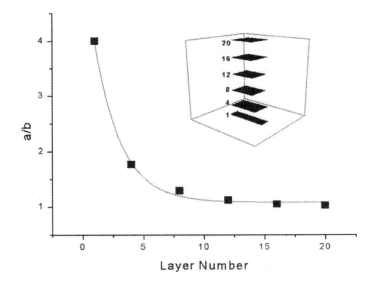

Fig. 14: Evolution of the aspect ratio a/b of base dimensions of 3D islands in a multilayer film of 20 bilayers simulated for a single island column. The solid line is a guide to the eye. The inset shows a schematic view of the island shape evolution.

value less than unity. This assumption, combined with the dependence of x_0 on W, leads to a critical island size toward which all islands will evolve [109].

The same mechanism also explains the approach toward square islands described above. In particular, when an anisotropic island is buried, it produces a tensile region that is also anisotropic. However, the aspect ratio of the tensile region, which determines that of the new island, is reduced relative to that of the original island [109]. Thus the island size approaches a more isotropic shape. Figure 14 shows the evolution of a single island column, in a 3D simulation of multilayer film growth, for which the island's initial aspect ratio is 4:1. After about 12 layers, the aspect ratio has evolved to nearly 1:1.

3.2.3 Multilayer engineering

Knowledge of SK growth in single, strained heteroepitaxial layers and multilayer ordering provides additional means of control of the properties of ensembles of coherent, 3D islands. Mukhametzhanov et al. [104] demonstrated how to control independently the size and density of InAs/GaAs islands in a bilayer. The first layer is intentionally grown with a low density of islands, approximately $3.7 \times 10^{10}/\text{cm}^2$. Next a spacer layer is grown, followed by a second layer. The amount of deposited material in the second layer is larger than that of the first layer. In the absence of the buried layer, this larger coverage results in an island density of $9.0 \times 10^{10}/cm^2$. Because of the strain inhomogeneity arising from the buried layer, the island density is only $4.8 \times 10^{10}/cm^2$. Furthermore, the island size distribution in the second layer is slightly narrower than that of the buried layer.

A common observation among samples of multilayer Ge/Si QD arrays is that, even with a constant amount of Ge deposited in each layer, the islands become larger with increasing layer number. This increase in size is primarily caused by a concomitant decrease in island density, as is evident from Figs. 11–12. However, careful study [110, 111] of multilayer structures reveals that the wetting-layer thickness decreases with increasing layer number, giving rise to an increase in average island size with increasing layer number even when the island density remains constant. For Ge/Si heterostructures with a 2.5 nm Si spacer layer, the measured wetting-layer thickness changes from 4 ML for the first Ge layer to about 1.66 ML in the second Ge layer [110]. In this experiment, sparse arrays of islands were used so that the island density was the same for all layers. In order to produce equal sized quantum dots among different layers, Thanh et al. compensated for this reduction in wetting-layer thickness by reducing the total Ge dose with increasing layer number [110].

3.3 Summary and Outlook

In this chapter we have described the spontaneous formation, or self-assembly, of 3D islands in strained, heteroepitaxial, semiconductor thin films. Such islands are dislocation-free, have lateral dimensions on the order of 10 nm, and form in a wide variety of systems, including Ge/Si and InAs/GaAs, making them promising candidates for semiconductor quantum dots. Coherent, 3D islands form via the Stranski-Krastanov growth mode, in which the deposited material initially wets the substrate, but forms islands after a system-dependent critical thickness. Strain can alter both the kinetics and equilibrium properties that govern island evolution, producing dramatic affects on island nucleation and coarsening phenomena, and also on the size and size distributions of islands within a strained layer. Research on semiconductor quantum dots has begun to focus on ways to manipulate and engineer 3D islands and island distributions for use in devices. Such methods include the use of substrate patterning, vicinal and step-bunched substrates, dislocation arrays, multilayers, and combinations of these techniques. Fundamental aspects of Stranski-Krastanov growth play a large role in developing such methods. We expect these developments to lead to increasing use of semiconductor quantum dots in novel and established micro- and optoelectronics technologies.

Acknowledgements

We thank Don Savage for many helpful discussions. Preparation of this review was supported by DARPA, ONR, and NSF.

References

1. S. Fafard, Z. R. Wasilewski, C. N. Allen, K. Hinzer, J. P. McCaffrey, and Y. Feng, "Lasing in quantum-dot ensembles with sharp adjustable electronic shells", *Appl. Phys. Lett.* **75** (1999) 986–988.
2. V. M. Ustinov, E. R. Weber, S. Ruvimov, Z. Liliental-Weber, A. E. Zhukov, A. Y. Egorov, A. R. Kovsh, A. F. Tsatsulńikov, and P. S. Kop'ev, "Effect of matrix on InAs self-organized quantum dots on InP substrate", *Appl. Phys. Lett.* **72** (1998) 362–364.
3. N. N. Ledentsov, V. A. Shchukin, M. Grundmann, N. Kirstaedter, J. Böhrer, O. Schmidt, D. Bimberg, V. M. Ustinov, A. Y. Egorov, A. E. Zhukov, P. S. Kop'ev, S. V. Zaitsev, N. Y. Gordeev, Z. I. Alferov, A. Borovkov, A. O. Kosogov, S. S. Ruvimov, P. Werner, U. Gösele, and J. Heydenreich, "Direct formation of vertically coupled quantum dots in Stranski-Krastanow growth", *Phys. Rev. B* **54** (1996) 8743–8750.
4. L. Friedman, G. Sun, and R. A. Soref, "SiGe/Si THz laser based on transitions between inverted mass light-hole and heavy-hole subbands", *Appl. Phys. Lett.* **78** (2001) 401–403.
5. T. Lundstrom, W. Schoenfeld, H. Lee, and P. M. Petroff, "Exiton storage in semiconductor self-assembled quantum dots", *Science* **286** (1999) 2312–2314.
6. M. C. Bodefeld, R. J. Warburton, K. Karrai, J. P. Kotthaus, G. Medeiros-Ribeiro, and P. M. Petroff, "Storage of electrons and holes in self-assembled InAs quantum dots", *Appl. Phys. Lett.* **74** (1999) 1839–1841.
7. P. Schittenhelm, C. Engel, F. Findeis, G. Abstreiter, A. A. Darhuber, G. Bauer, A. O. Kosogov, and P. Werner, "Self-assembled Ge dots: Growth, characterization, ordering, and applications", *J. Vac. Sci. Tech. B* **16** (1998) 1575–1581.
8. R. Soref, "Applications of silicon-based optoelectronics", *MRS Bulletin* **23** (1998) 20–24.
9. N. H. Bonadeo, J. Erland, D. Gammon, D. Park, D. S. Katzer, and D. G. Steel, "Coherent optical control of the quantum state of a single quantum dot", *Science* **282** (1998) 1473–1476.
10. L. Tsybeskov, G. F. Grom, P. M. Fauchet, J. P. McMaffrey, J. M. Baribeau, G. I. Sproule, and D. J. Lockwood, "Phonon-assisted tunneling and interface quality in nanocrystalline Si/amorphous SiO_2 superlattices", *Appl. Phys. Lett.* **75** (1999) 2265–2267.
11. S. Chan and P. M. Fauchet, "Tunable, narrow, and directional luminescence from porous silicon light emitting devices", *Appl. Phys. Lett.* **75** (1999) 274–276.
12. C. J. Kiely, J. Fink, J. G. Zheng, M. Brust, D. Bethell, and D. J. Schiffrin, "Ordered colloidal nanoalloys", *Adv. Mat.* **12** (2000) 640–643.
13. G. Schmid, "Large clusters and colloids. Metals in the embryonic state", *Chem. Rev.* **92** (1992) 1709–1727.
14. J. Legrand, C. Petit, D. Bazin, and M. P. Pileni, "Collective effect on magnetic properties of 2D superlattices of nanosized cobalt particles", *Appl. Surf. Sci.* **164** (2000) 186–192.
15. D. C. Ralph, S. Gueron, C. T. Black, and M. Tinkham, "Electron energy levels in superconducting and magnetic nanoparticles", *Physica B* **280** (2000) 420–424.
16. V. I. Klimov, A. A. Mikhailovsky, S. Xu, A. Malko, J. A. Hollingsworth, C. A. Leatherdale, H.-J. Eisler, and M. G. Bawendi, "Optical gain and stimulated emission in nanocrystal quantum dots", *Science* **290** (2000) 314–317.
17. D. J. Eaglesham and M. Cerullo, "Dislocation-free Stranski-Krastanow growth of Ge on Si(100)", *Phys. Rev. Lett.* **64** (1990) 1943–1946.
18. Y. W. Mo, D. E. Savage, B. S. Swartzentruber, and M. G. Lagally, "Kinetic pathway in Stranski-Krastanov growth of Ge on Si(001)", *Phys. Rev. Lett.* **65** (1990) 1020–1023.

19. G. Abstreiter, P. Schittenhelm, C. Engel, E. Silveira, A. Zrenner, and D. Meertens, "Growth and characterization of self-assembled Ge-rich islands on Si", *Sem. Sci. Tech.* **11** (1996) 1521–1528.
20. G. Jin, J. L. Liu, S. G. Thomas, Y. H. Luo, K. L. Wang, and B. Y. Nguyen, "Controlled arrangement of self-organized Ge islands on patterned Si (001) substrates", *Appl. Phys. Lett.* **75** (1999) 2752–2754.
21. F. Flack, N. Samarth, V. Nikitin, P. A. Crowell, J. Shi, J. Levy, and D. D. Awschalom, "Near-field optical spectroscopy of localized excitons in strained CdSe quantum dots", *Phys. Rev. B* **54** (1996) R17312–R17315.
22. S. H. Xin, P. D. Wang, A. Yin, C. Kim, M. Dobrowolska, J. L. Merz, and J. K. Furdyna, "Formation of self-assembling CdSe quantum dots on ZnSe by molecular beam epitaxy", *Appl. Phys. Lett.* **69** (1996) 3884–3886.
23. K. Kitamura, H. Umeya, A. Jia, M. Shimotomai, Y. Kato, M. Kobayashi, A. Yoshikawa, and K. Takahashi, "Self-assembled CdS quantum-dot structures grown on ZnSe and ZnSSe", *J. Cryst. Growth* **214** (2000) 680–683.
24. G. Springholz, V. Holy, M. Pinczolits, and G. Bauer, "Self-organized growth of three-dimensional quantum-dot crystals with fcc-like stacking and a tunable lattice constant", *Science* **282** (1998) 734–737.
25. M. Pinczolits, G. Springholz, and G. Bauer, "Direct formation of self-assembled quantum dots under tensile strain by heteroepitaxy of PbSe on PbTe(111)", *Appl. Phys. Lett.* **73** (1998) 250–252.
26. V. Bressler-Hill, S. Varma, A. Lorke, B. Z. Nosho, P. M. Petroff, and W. H. Weinberg, "Island scaling in strained heteroepitaxy: InAs/GaAs(001)", *Phys. Rev. Lett.* **74** (1995) 3209–3212.
27. X. Qianghua, A. Madhukar, P. Chen, and N. P. Kobayashi, "Vertically self-organized InAs quantum box islands on GaAs(100)", *Phys. Rev. Lett.* **75** (1995) 2542–2545.
28. R. Leon, S. Fafard, D. Leonard, J. L. Merz, and P. M. Petroff, "Visible luminescence from semiconductor quantum dots in large ensembles", *Appl. Phys. Lett.* **67** (1995) 521–523.
29. P. Ramvall, S. Tanaka, S. Nomura, P. Riblet, and Y. Aoyagi, "Confinement induced decrease of the exciton-longitudinal optical phonon coupling in GaN quantum dots", *Appl. Phys. Lett.* **75** (1999) 1935–1937.
30. B. Damilano, N. Grandjean, F. Semond, J. Massies, and M. Leroux, "From visible to white light emission by GaN quantum dots on Si(111) substrate", *Appl. Phys. Lett.* **75** (1999) 962–964.
31. K. Tachibana, T. Someya, S. Ishida, and Y. Arakawa, "Selective growth of InGaN quantum dot structures and their microphotoluminescence at room temperature", *Appl. Phys. Lett.* **76** (2000) 3212–3214.
32. N. Carlsson, W. Seifert, A. Petersson, P. Castrillo, M. E. Pistol, and L. Samuelson, "Study of the two-dimensional-three-dimensional growth mode transition in metalorganic vapor phase epitaxy of GaInP/InP quantum-sized structures", *Appl. Phys. Lett.* **65** (1994) 3093–3095.
33. A. Kurtenbach, K. Eberl, and T. Shitara, "Response to 'Comment on "Nanoscale InP islands embedded in InGaP"'", *Appl. Phys. Lett.* **67** (1995) 1168–1169.
34. M. K. Zundel, A. P. Specht, K. Eberl, N. Y. Jin-Phillip, and F. Phillipp, "Structural and optical properties of vertically aligned InP quantum dots", *Appl. Phys. Lett.* **71** (1997) 2972–2974.
35. F. Hatami, N. N. Ledentsov, M. Grundmann, J. Bohrer, F. Heinrichsdorff, M. Beer, D. Bimberg, S. S. Ruvimov, P. Werner, U. Gösele, J. Heydenreich, U. Richter, S. V. Ivanov, B. Y. Meltser, P. S. Kop'ev, and Z. I. Alferov, "Radiative recombination in type-II GaSb/GaAs quantum dots", *Appl. Phys. Lett.* **67** (1995) 656–658.

36. I. Daruka, A. L. Barabasi, Y. Chen, and J. Washburn, "Island formation and critical thickness in heteroepitaxy [Comment and reply]", *Phys. Rev. Lett.* **78** (1997) 3027–3028.
37. E. Bauer, "Phänomenologische Theorie der Kristallabscheidung an Oberflächen", *Z. Kristallogr.* **110** (1958) 372–394.
38. F. C. Frank and J. H. van der Merwe, "One-dimensional dislocations", *Proc. Roy. Soc. London A* **198** (1949) 205–225.
39. M. Volmer and A. Weber, "Keimbildung in Übersättigten Gebilden", *Z. Phys. Chem* **119** (1926) 277.
40. I. Stranski and L. Krastanow, "Zur Theorie der Orientierten Ausscheidung von Ionenkristallen Aufeinander", *Sitz. Ber. Akad. Wiss. Wein.* **146** (1938) 797–810.
41. E. Bauer and J. H. van der Merwe, "Structure and growth of crystalline superlattices: From monolayer to superlattice", *Phys. Rev. B* **33** (1986) 3657–3671.
42. R. Bruinsma and A. Zangwill, "Morphological transitions in solid epitaxial overlayers", *Europhys. Lett.* **4** (1987) 729–735.
43. S. Guha, A. Madhukar, and K. C. Rajkumar, "Onset of incoherency and defect introduction in the initial stages of molecular beam epitaxial growth of highly strained In_xGa_{1-x}/As on GaAs(100)", *Appl. Phys. Lett.* **57** (1990) 2110–2112.
44. A. A. Williams, J. M. C. Thornton, J. E. Macdonald, R. G. van Silfhout, J. F. van der Veen, M. S. Finney, A. Johnson, and C. Norris, "Strain relaxation during the initial stages of growth in Ge/Si(001)", *Phys. Rev. B* **43** (1991) 5001–5011.
45. J. A. Floro, E. Chason, S. R. Lee, R. D. Twesten, R. Q. Hwang, and L. B. Freund, "Real-time stress evolution during $Si_{1-x}Ge_x$ heteroepitaxy: dislocations, islanding, and segregation", *J. Ele. Mat.* **26** (1997) 969–979.
46. J. A. Floro, E. Chason, L. B. Freund, R. D. Twesten, R. Q. Hwang, and G. A. Lucadamo, "Evolution of coherent islands in $Si_{1-x}Ge_x$/Si(001)", *Phys. Rev. B* **59** (1999) 1990–1998.
47. J. Tersoff and F. K. LeGoues, "Competing relaxation mechanisms in strained layers", *Phys. Rev. Lett.* **72** (1994) 3570–3573.
48. V. A. Shchukin and D. Bimberg, "Spontaneous ordering of nanostructures on crystal surfaces", *Rev. Mod. Phys.* **71** (1999) 1125–1171.
49. G. Medeiros-Ribeiro, A. M. Bratkovski, T. I. Kamins, D. A. A. Ohlberg, and R. S. Williams, "Shape transition of germanium nanocrystals on a silicon (001) surface from pyramids to domes", *Science* **279** (1998) 353–355.
50. Y. Nabetani, T. Ishikawa, S. Noda, and A. Sasaki, "Initial growth stage and optical properties of a three-demensional InAs structure on GaAs", *J. Appl. Phys.* **76** (1994) 347–351.
51. J. Márquez, L. Geelhaar, and K. Jacobi, "Atomically resolved structure of InAs quantum dots", *Appl. Phys. Lett.* **78** (2001) 2309–2311.
52. H. Röder, K. Bromann, H. Brune, and K. Kern, "Strain mediated two-dimensional growth kinetics in metal heteroepitaxy: Ag/Pt(111)", *Surf. Sci.* **376** (1997) 13–31.
53. H. A. van der Vegt, H. M. van Pinxteren, M. Lohmeier, E. Vlieg, and J. M. C. Thornton, "Surfactant-induced layer-by-layer growth of Ag on Ag(111)", *Phys. Rev. Lett.* **68** (1992) 3335–3338.
54. D. Vanderbilt and L. K. Wickham, in *Evolution of Thin-Film and Surface Microstructure*, Vol. 202 of *MRS Proceedings*, edited by C. V. Thompson, J. Y. Tsao, and D. J. Srolovitz (MRS, Pittsburgh, 1991), pp. 555–560.
55. C. Ratsch and A. Zangwill, "Equilibrium theory of the Stranski-Krastanov epitaxial morphology", *Surf. Sci.* **293** (1993) 123–131.
56. F. M. Ross, J. Tersoff, and R. M. Tromp, "Coarsening of self-assembled Ge quantum dots on Si(001)", *Phys. Rev. Lett.* **80** (1998) 984–987.

57. I. Daruka and A. L. Barabasi, "Dislocation-free island formation in heteroepitaxial growth: a study at equilibrium", *Phys. Rev. Lett.* **79** (1997) 3708–3711.
58. I. Daruka and A. L. Barabasi, "Equilibrium phase diagrams for dislocation free self-assembled quantum dots", *Appl. Phys. Lett.* **72** (1998) 2102–2104.
59. V. A. Shchukin, N. N. Ledentsov, P. S. Kop'ev, and D. Bimberg, "Spontaneous odering of arrays of coherent strained islands", *Phys. Rev. Lett.* **75** (1995) 2968–2971.
60. R. J. Asaro and W. A. Tiller, "Interface morphology development during stress corrosion cracking: Part I. via surface diffusion", *Met. Trans.* **3** (1972) 1789–1796.
61. M. Grinfeld, "Two-dimensional islanding atop stressed solid helium and epitaxial films", *Phys. Rev. B* **49** (1994) 8310–8319.
62. A. J. Pidduck, D. J. Robbins, and A. Cullis, in *Microscopy of Semiconducting Materials*, edited by A. G. Cullis, J. L. Hutchinson, and A. E. Staton-Bevan (IOP, Bristol, UK, 1993), pp. 609–612.
63. K. M. Chen, D. E. Jesson, S. J. Pennycook, T. Thundat, and R. J. Warmack, "Critical nuclei shapes in the stress-driven 2D-to-3D transition", *Phys. Rev. B* **56** (1997) R1700–R1703.
64. P. Sutter and M. G. Lagally, "Nucleationless three-dimensional island formation in low-misfit heteroepitaxy", *Phys. Rev. Lett.* **84** (2000) 4637–4640.
65. R. M. Tromp, E. M. Ross, and M. C. Reuter, "Instability-driven SiGe island growth", *Phys. Rev. Lett.* **84** (2000) 4641–4644.
66. M. G. Lagally, "An atomic-level view of kinetic and thermodynamic influences in the growth of thin films", *Jpn. J. Appl. Phys.* **32** (1993) 1493–1501.
67. F. Wu, "Scanning tunneling microscopy studies of growth of Si and Ge on Si(001)", Ph.D. thesis, University of Wisconsin–Madison, 1996.
68. A. Vailionis, B. Cho, G. Glass, P. Desjardins, D. G. Cahill, and J. E. Greene, "Pathway for the strain-driven two-dimentional to three-dimensional transition during growth of Ge on Si(001)", *Phys. Rev. Lett.* **85** (2000) 3672–3675.
69. D. E. Jesson, M. Kastner, and B. Voigtländer, "Direct observation of subcritical fluctuations during the formation of strained semiconductor islands", *Phys. Rev. Lett.* **84** (2000) 330–333.
70. B. Lewis and J. C. Anderson, *Nucleation and Growth of Thin Films* (Academic Press, New York, 1978).
71. M. Kästner and B. Voigtländer, "Kinetically self-limiting growth of Ge islands on Si(001)", *Phys. Rev. Lett.* **82** (1999) 2745–2748.
72. F. K. LeGoues, M. C. Reuter, J. Tersoff, M. Hammar, and R. M. Tromp, "Cyclic growth of strain-relaxed islands", *Phys. Rev. Lett.* **73** (1994) 300–303.
73. N. P. Kobayashi, T. R. Ramachandran, P. Chen, and A. Madhukar, "*In situ*, atomic force microscope studies of the evolution of InAs three-dimensional islands on GaAs(001)", *Appl. Phys. Lett.* **68** (1996) 3299–3301.
74. D. Leonard, K. Pond, and P. M. Petroff, "Critical layer thickness for self-assembled InAs islands on GaAs", *Phys. Rev. B* **50** (1994) 11687–11692.
75. A. Ponchet, A. Le Corre, H. L'Haridon, B. Lambert, and S. Salaün, "Relationship between self-organization and size of InAs islands on InP(001) grown by gas-source molecular beam epitaxy", *Appl. Phys. Lett.* **67** (1995) 1850–1852.
76. Y. Chen and J. Washburn, "Structural transition in large-lattice-mismatch heteroepitaxy", *Phys. Rev. Lett.* **77** (1996) 4046–4049.
77. D. E. Jesson, G. Chen, K. M. Chen, and S. J. Pennycook, "Self-limiting growth of strained faceted islands", *Phys. Rev. Lett.* **80** (1998) 5156–5159.
78. G. Medeiros-Ribeiro, T. I. Kamins, D. A. A. Ohlberg, and R. S. Williams, "Annealing of Ge nanocrystals on Si(001) at 550 degrees C: Metastability of huts and the stability of pyramids and domes", *Phys. Rev. B* **58** (1998) 3533–3536.

79. J. A. Floro, G. A. Lucadamo, E. Chason, L. B. Freund, M. Sinclair, R. D. Twesten, and R. Q. Hwang, "SiGe island shape transitions induced by elastic repulsion", *Phys. Rev. Lett.* **80** (1998) 4717–4720.
80. C. M. Reaves, R. I. Pelzel, G. C. Hsueh, W. H. Weinberg, and S. P. DenBaars, "Formation of self-assembled InP islands on a GaInP/GaAs(311)A surface", *Appl. Phys. Lett.* **69** (1996) 3878–3880.
81. F. M. Ross, R. M. Tromp, and M. C. Reuter, "Transition States Between Pyramids and Domes During Ge/Si Island Growth", *Science* **286** (1999) 1931–1934.
82. I. Daruka, J. Tersoff, and A. L. Barabasi, "Shape transition in growth of strained islands", *Phys. Rev. Lett.* **82** (1999) 2753–2756.
83. J. Drucker and S. Chaparro, "Diffusional narrowing of Ge on Si(100) coherent island quantum dot size distributions", *Appl. Phys. Lett.* **71** (1997) 614–616.
84. J. S. Sullivan, E. Mateeva, H. Evans, D. E. Savage, and M. G. Lagally, "Properties of $Si_{1-x}Ge_x$ three dimensional islands", *J. Vac. Sci. Technol. A* **17** (1999) 2345–2350.
85. T. R. Ramachandran, A. Madhukar, I. Mukhametzhanov, R. Heitz, A. Kalburge, Q. Xie, and P. Chen, "Nature of Stranski-Krastanow growth of InAs on GaAs(001)", *J. Vac. Sci. Tech. B* **16** (1998) 1330–1333.
86. J. S. Sullivan, "The Stranski-Krastanov growth mode of $Si_{1-x}Ge_x$ on Si(001)", Ph.D. thesis, University of Wisconsin–Madison, 1999.
87. C. Hernandez, Y. Campidelli, D. Simon, D. Bensahel, I. Sagnes, G. B. Patriarche, and S. Sauvage, "Ge/Si self-assembled quantum dots grown on Si(001) in an industrial high-pressure chemical vapor deposition reactor", *J. Appl. Phys.* **86** (1999) 1145–1148.
88. G. Jin, J. Wan, Y. H. Luo, J. L. Liu, and K. L. Wang, "Uniform and ordered self-assembled Ge dots on patterned Si substrates with selectively epitaxial growth technique", *J. Crys. Growth.* **227–228** (2001) 1100–1105.
89. T. I. Kamins, R. S. Williams, and D. P. Basile, "Self-aligning of self-assembled Ge islands on Si(001)", *Nanotechnology* **10** (1999) 117–121.
90. T. I. Kamins, D. A. A. Ohlberg, R. S. Williams, W. Zhang, and S. Y. Chou, "Positioning of self-assembled, single-crystal, germanium islands by silicon nanoimprinting", *Appl. Phys. Lett.* **74** (1999) 1773–1775.
91. T. I. Kamins and R. S. Williams, "Lithographic positioning of self-assembled Ge islands on Si(001)", *Appl. Phys. Lett.* **71** (1997) 1201–1203.
92. K. H. Ploog and R. Nötzel, "Uniform III-V semiconductor quantum wire and quantum dot arrays by natural self-faceting on patterned substrates", *Thin Solid Films* **367** (2000) 32–39.
93. P. Rugheimer, E. Mateeva, D. E. Savage, and M. G. Lagally, "Novel morphology and kinetics of Ge deposited on SOI substrates", *To be published* (2001).
94. C. Teichert, J. C. Bean, and M. G. Lagally, "Self-organized nanostructures in $Si_{1-x}Ge_x$ films on Si(100)", *Appl. Phys. A* **67** (1998) 675–685.
95. C. Teichert, J. Barthel, H. P. Oepen, and J. Kirschner, "Fabrication of nanomagnet arrays by shadow deposition on self-organized semiconductor substrates", *Appl. Phys. Lett.* **74** (1999) 588–590.
96. H. Omi and T. Ogino, "Self-organization of Ge islands on high-index Si substrates", *Phys. Rev. B* **59** (1999) 7521–7528.
97. H. Omi and T. Ogino, "Positioning of self-assembling Ge islands on Si(111) mesas by using atomic steps", *Thin Solid Films* **369** (2000) 88–91.
98. C. Teichert, C. Hofer, K. Lyutovich, M. Bauer, and E. Kasper, "Interplay of dislocation network and island arrangement in SiGe films grown on Si(100)", *Thin Solid Films* **380** (2000) 25–28.
99. A. Konkar, R. Heitz, T. R. Ramachandran, P. Chen, and A. Madhukar, "Fabrication of strained InAs island ensembles on nonplanar patterned GaAs(001) substrates", *J.*

Vac. Sci. Tech. B **16** (1998) 1334–1338.
100. J. Tersoff, C. Teichert, and M. G. Lagally, "Self-organization in growth of quantum dot superlattices", *Phys. Rev. Lett.* **76** (1996) 1675–1678.
101. C. Teichert, M. G. Lagally, L. J. Peticolas, J. C. Bean, and J. Tersoff, "Stress-induced self-organization of nanoscale structures in SiGe/Si multilayer films", *Phys. Rev. B* **53** (1996) 16334–16337.
102. E. Mateeva, P. Sutter, J. C. Bean, and M. G. Lagally, "Mechanism of organization of three-dimensional islands in SiGe/Si multilayers", *Appl. Phys. Lett.* **71** (1997) 3233–3235.
103. A. A. Darhuber, P. Schittenhelm, V. Holy, J. Stangl, G. Bauer, and G. Abstreiter, "High-resolution x-ray diffraction from multilayered self-assembled Ge dots", *Phys. Rev. B* **55** (1997) 15652–15663.
104. I. Mukhametzhanov, R. Heitz, J. Zeng, P. Chen, and A. Madhukar, "Independent manipulation of density and size of stress-driven self-assembled quantum dots", *Appl. Phys. Lett.* **73** (1998) 1841–1843.
105. O. Kienzle, F. Ernst, M. Rühle, O. G. Schmidt, and K. Eberl, "Germanium 'quantum dots' embedded in silicon: Quantitative study of self-alignment and coarsening", *Appl. Phys. Lett.* **74** (1999) 269–271.
106. T. S. Kuan and S. S. Iyer, "Strain relaxation and ordering in SiGe layers grown on (100), (111), and (110) Si surfaces by molecular-beam epitaxy", *Appl. Phys. Lett.* **59** (1991) 2242–2244.
107. J. Y. Yao, T. G. Andersson, and G. L. Dunlop, "The interfacial morphology of strained epitaxial $In_xGa_{1-x}As/GaAs$", *J. Appl. Phys.* **69** (1991) 2224–2230.
108. V. A. Shchukin, D. Bimberg, V. G. Malyshkin, and N. N. Ledentsov, "Vertical correlations and anticorrelations in multisheet arrays of two-dimensional islands", *Phys. Rev. B* **57** (1998) 12262–12274.
109. F. Liu, S. E. Davenport, H. M. Evans, and M. G. Lagally, "Self-organized replication of 3D coherent island size and shape in multilayer heteroepitaxial films", *Phys. Rev. Lett.* **82** (1999) 2528–2531.
110. V. L. Thanh, V. Yam, P. Boucaud, Y. Zheng, and D. Bouchier, "Strain-driven modification of the Ge/Si growth mode in stacked layers: a way to produce Ge islands having equal size in all layers", *Thin Solid Films* **369** (2000) 43–48.
111. O. G. Schmidt, O. Kienzle, Y. Hao, K. Eberl, and F. Ernst, "Modified Stranski-Krastanov growth in stacked layers of self-assembled islands", *Appl. Phys. Lett.* **74** (1999) 1272–1274.

GROWTH, STRUCTURES, AND OPTICAL PROPERTIES OF III-NITRIDE QUANTUM DOTS

DAMING HUANG,[a] MICHAEL A. RESHCHIKOV, and HADIS MORKOÇ

Virginia Commonwealth University
Department of Electrical Engineering and Physics Department
Richmond, VA 23284, USA

This article reviews the advances in the growth of III-nitride quantum dots achieved in the last few years and their unique properties. The growth techniques and the structural and optical properties associated with quantum confinement, strain, and polarization in $GaN/Al_xGa_{1-x}N$ and $In_xGa_{1-x}N/GaN$ quantum dots are discussed in detail.

1. Introduction

Investigations of semiconductor quantum dots (QDs) have been very extensive, particularly in the last decade.[1,2] These are fuelled by unique physical phenomena and potential device applications. As compared to bulk (three-dimensional or 3D) materials and quantum well (QW) (two-dimensional or 2D) structures, QD is the prototype of zero-dimensional system and possesses many unique properties. The electronic states in a QD are spatially localized and the energy is fully quantized, similar to a single atom or atomic system. So the system is more stable against any thermal perturbation. In addition, due to the quantization, the electronic density of states near the band gap is higher than in 3D and 2D systems, leading to a higher probability for optical transitions. Furthermore, the electron localization may dramatically reduce the scattering of electrons by bulk defects and reduce the rate of non-radiative recombination. These properties, among the others, are directly relevant to the high thermal stability and high quantum efficiency in light emitting and detecting devices, and are of great importance in terms of device applications.

III-nitride semiconductors have also been investigated very extensively in the last decade.[3,4] GaN is a direct and wide band-gap semiconductor and when alloyed with InN and AlN, a spectrum from visible to ultraviolet can be covered. GaN has large high-field electron velocity, large breakdown field, large thermal conductivity, and robustness. It is also a strongly polar crystal as compared to other III-V semiconductors, and thus shows a large piezoelectric effect. In the last few years, GaN based light emitting devices (LEDs) and laser diodes (LDs) with green, blue, and violet colors have been achieved using thin films or QWs.[5] A flurry of interest in low-dimensional GaN and other III-nitrides is in part associated with the desire to develop new optoelectronic devices with improved quality and wider applications. Development of the light emitters with QDs, for example, is expected to have a lower threshold current in LDs and a higher thermal stability of the devices.[6]

To attain the advantages of the QDs, many requirements must be met in material preparation. The most important one is the size and of its distribution of QDs. Depending

---

[a] Also with: Physics Department, Fudan University, Shanghai 200433, China.

on the material and the dot shape, the maximum size should be near or less than some characteristic length of the electrons in the bulk III-nitride such as exciton radius. The typical value is on the order of a few nm. With such a small size, the practical applications are thus often associated with a large assembly of QDs rather than a single one. This implies that the size uniformity of the dot assembly is critical. The fluctuation in dot size produces an inhomogeneous broadening in quantized energy levels and may destroy the very properties expected from a single QD. In addition to the size uniformity, the spatial position of each QD is also important in many applications. The random rather than well-ordered distribution may damage the coherence of the optical and electronic waves propagating through the system. Similar to other semiconductor heterostructures, the surface or interface of the QDs must also be free of defects. Otherwise, the surface/interface may become the effective scattering centers to electrons. Fabrication of QD assembly with small and uniform size, high density, well-ordered placement, and defect-free materials remains as great challenges today in any semiconductor system, especially in III-Nitride materials.

The properties of III-nitride QDs are closely related to those of bulk materials that have been reviewed in many articles.[3,4] In Table 1 we list those material and physical parameters of GaN, AlN, and InN relevant to the QD investigations. Although bulk GaN, AlN, and InN can all crystallize in wurtzite, zincblende, and rocksalt structures, in ambient conditions, the thermodynamically stable structure, however, is the wurtzitic phase. In this case, the GaN, AlN, and InN have direct room temperature band-gaps of 3.4, 6.2, and 1.9 eV, respectively. The band gap, as well as the detailed electronic structure, can be modified by incorporating alloy and low-dimensional structures such as QWs and QDs. Depending on material, size, and shape, very rich electronic and optical phenomena can be observed from quantum structures that are not expected in bulk.

III-nitride QDs are commonly the strained systems. As listed in Table 1, the lattice constants for bulk wurtzitic GaN are 3.189 Å and 5.185 Å in in-plane and c directions respectively. For AlN, they are 3.11 Å and 4.98 Å. For InN, the values are 3.55 Å and 5.76 Å. The lattice constants of $In_xGa_{1-x}N$ and $Al_xGa_{1-x}N$ alloys can be roughly estimated by assuming linear interpolations from two ending compounds. Using these data, the lattice mismatch between GaN and AlN in basal plane is 2.5% and is much larger between GaN and InN (10%). The lattice mismatch and its induced strain have profound effect on the growth, structures, and properties of semiconductor QDs.

Table 1. Material parameters of wurtzite III-nitride semiconductors.[3,4]

Parameter	Notation	Unit	GaN	AlN	InN
Lattice Constant	a	Å	3.189	3.112	3.548
Lattice Constant	c	Å	5.185	4.982	5.760
Thermal Coefficient	$\Delta a/a$	K^{-1}	5.59×10^{-6}	4.2×10^{-6}	
Thermal Coefficient	$\Delta c/c$	K^{-1}	3.17×10^{-6}	5.3×10^{-6}	
Band Gap, 300K	E_g	eV	3.42	6.2	1.89
Band Gap, 4K	E_g	eV	3.505	6.28	
Electron Effective Mass	m_e	m_0	0.22		
Hole Effective Mass	m_h	m_0	>0.8		
Elastic Constant	C_{13}	GPa	94	127	100
Elastic Constant	C_{33}	GPa	390	382	392
Static Dielectric Constant	ε_r	ε_0	10.4	8.5	15.3
Spontaneous Polarization	P_{spon}	C/m^2	-0.029	-0.081	-0.032
Piezoelectric Coefficient	e_{31}	C/m^2	-0.49	-0.60	-0.57
Piezoelectric Coefficient	e_{33}	C/m^2	0.73	1.46	0.97
Binding Energy, Exciton A	E_{xb}	meV	21		

Wurtzite III-nitrides are noncentrosymmetric and uniaxial crystals. They exhibit large effect of spontaneous and strain induced polarization. The piezoelectric coefficients, as listed in Table 1, are almost an order of magnitude larger than in traditional III-V compounds such as GaAs. The large amount of polarization charge may appear at the heterointerface even in the absence of strain due to the difference in the spontaneous polarization across the interface. The very strong internal electric field induced by the polarization charge and piezoelectricity is very unique to III-nitride heterostructures and has a dramatic effect on the properties of QDs.

The purpose of this article is to review the advances in the growth of III-nitride QDs achieved in the last few years and their unique properties. Issues relevant to the growth and the structures of the QDs will be described in the next section. The properties associated with quantum confinement, strain, and polarization in $GaN/Al_xGa_{1-x}N$ and $In_xGa_{1-x}N/GaN$ QDs will be discussed in detail in Section 3.

2. Growth and Structures

The great majority of III-nitride QDs are grown by molecular beam epitaxy (MBE) and metalorganic chemical vapor deposition (MOCVD). Due to the lack of a suitable material of both lattice and thermally matched to GaN, III-nitride heterostructures are commonly grown on foreign substrates.[7] Sapphire (α-Al_2O_3) is the one most extensively used since the high quality GaN has been grown and inexpensive wafers are available up to 6 inches in diameter, though 3 inch varieties are common for epitaxial growth. Wurtzitic GaN QDs have also been grown on other substrates such as 6H-SiC(0001) and Si(111). In addition, the cubic GaN QDs has been grown on cubic substrates such as 3C-SiC (001).

There are large mismatches of both lattice constant and thermal expansion coefficient between GaN and sapphire. The surface polarity of both materials is also very different. For the growth of high quality crystal, a fully relaxed buffer layer is necessary to isolate the epilayers from the deleterious effect of the substrate. In the case of MBE, an AlN buffer layer is commonly used and shown to lead to better epilayer quality. For MOCVD, either GaN or AlN can be used as buffer layer. In some cases, thick GaN or AlN layer grown on sapphire by MOCVD and GaN by HVPE are used as the templates for further MBE growth.

The epitaxial growth by MBE and MOCVD is essentially a non-equilibrium process. However, it is very useful to categorize it into three different modes as in the equilibrium theory. As schematically shown in Figure 1,[8,9] Frank-Van der Merwe mode represents a layer-by-layer or 2D growth. Volmer-Weber mode corresponds to island or 3D growth.

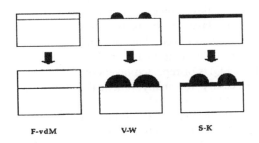

Fig. 1. Schematic diagram of the three possible growth modes: Frank-van der Merwe, Volmer-Weber, and Stranski-Krastanow. (Ref. 9)

Stranski-Krastanow (SK) mode is 2D growth of a few monolayers, called as wetting layer, followed by 3D island formation. The last mode is the one most relevant to the growth of semiconductor QDs. The investigation of InAs QDs grown on GaAs shows that the 2D to 3D transition in SK mode is in fact a first order phase transition.[10]

Experimentally, the growth mode depends not only on the materials of both epilayer and substrate but also on the growth conditions such as substrate temperature and flow-rates of various sources. Essentially, it is the result of competition between the kinetic energy of adatoms and the free energies of bulk, surfaces and interface. For a lattice-matched system, in the limit of equilibrium growth, the layer-by-layer growth is favored if the energy of substrate surface is higher than the sum of the epilayer surface and interface energy. Island or 3D growth can be realized by changing the surface and interface energy. In a lattice-mismatched system, the bulk elastic energy induced by the strain in the epilayer plays an important role. Since it increases with the layer thickness, a strain relaxation is expected when the layer thickness is increased beyond a critical value. In fact, the SK mode is mostly observed when an epilayer is subject to a compressive strain. In this case, the stress field tends to force the adatoms to coalesce. The strain energy can be partially released by the formation of islands through elastic relaxation, without any dislocations in the islands.[8] When an epilayer is subject to a tensile strain, the growth usually continues to be 2D and the strain energy is released through plastic relaxation with the creation of dislocations. However, the transition from 2D to 3D growth is possible in the case of tensile strain if the lattice mismatch between an epilayer and substrate is large.[11] The spontaneous growth of QDs by either 3D or SK mode is known as the self-organized or self-assembled growth.

The self-assembled growth of GaN and $In_xGa_{1-x}N$ QDs have been demonstrated using both MBE and MOCVD techniques. The other growth methods such as laser ablation and reactive radio-frequency (rf) sputtering were also reported. In the following subsections, we will describe in detail the growth and structures of GaN and $In_xGa_{1-x}N$ QDs by MBE, MOCVD, and other techniques separately.

2.1 MBE

In III-nitride MBE growth, solid Ga, Al, and In are used as the III growth sources. Both radio-frequency (rf) nitrogen plasma[12,13,14,15,16,17] and ammonia (NH_3)[18,19] have been used as the nitrogen sources. The main steps in III-Nitride QDs growth on sapphire substrates are the substrate cleaning, substrate nitridation, buffer layer growth, and active layer growth. Substrate cleaning is a standard procedure and includes the steps of surface degreasing, chemical etching, and thermal annealing after introduction into the growth chamber.[15] Nitridation was achieved by exposing the sapphire surface to nitrogen plasma. During this step, a reactive layer most likely $Al_{1-x}O_xN$ may be formed. An AlN or AlN/GaN buffer layer is then followed. This buffer layer normally leads to an epilayer with Ga-face with a better quality than that with the N-face.[20,21] The active layer can be either a single layer of GaN QDs or repeated layers of GaN QDs separated by AlN spacer layers. The latter is necessary for many practical applications requiring higher QD density. In this case, the AlN is not only used to isolate adjacent QD layer, but also provide a flat surface for the growth of the following QD layer. The top layer of QDs may or may not be capped by AlN depending on the measurement to be performed.

Extensive investigation of GaN QDs grown by MBE has been carried out by Daudin and coworkers.[12,13,14,15,16,22,23] Typically, after substrate nitridation, a thin (~10-30 nm)

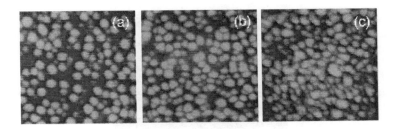

Fig. 2. 200nm×200nm AFM images from GaN quantum dots grown on AlN surface at (a) 725°C, (b) 705°C, (c) 685°C. The growth follows SK mode. The dot density decreases with growth temperature. (Ref. 14)

AlN buffer layer was grown at a temperature between 500-550 °C, followed by a thick (0.2-1.5 μm) AlN layer grown at higher temperature of 650-730 °C.[23] Sometimes the growth of the thick AlN layer was preceded by a thick (~2μm) GaN buffer layer.[12] The GaN QDs were grown on AlN by depositing 2-4 monolayers (MLs) of GaN at temperatures ranging from 680 to 730 °C. Due to the 2.5% lattice mismatch between GaN and AlN, under the growth conditions that were used, the growth follows a SK mode. After the 2D growth of a GaN wetting layer (about 2 MLs), the 3D growth is followed and the GaN QDs are formed.[12] It was found that the growth mode is sensitive to the substrate temperature. At growth temperature below 620 °C, the growth was purely two-dimensional (2D). Only at the elevated temperatures (680-730 °C), the growth transitions from 2D to 3D, i.e., the SK mode took place.[12]

The self-assembled GaN QDs have a disk-like shape, or more accurately, a truncated pyramid with a hexagonal base, with a base diameter a few times larger than the height. The dot size and density depend on growth condition, deposition time, as well as post-growth treatment. Atomic force microscopy (AFM) has been widely used to image the general morphology of the QDs that are not covered by any capping layer. Figure 2 shows the typical AFM images of the GaN QDs grown on AlN at three different temperatures near 700°C.[14] The dot density is higher than 10^{11} cm^{-2} and decreases with growth temperature. The density can be effectively reduced through a post-growth reorganization, called the ripening effect,[12] after GaN growth is finished. During this period, the sample was exposed to N plasma and kept at high temperature for approximately 50 seconds. Figure 3 clearly demonstrates the ripening effect. In this

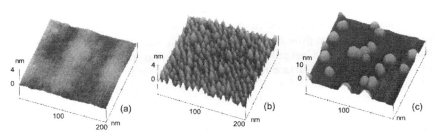

Fig. 3. (a) AFM image of smooth AlN surface. (b) GaN quantum dots formed by depositing the equivalent of four GaN monolayers on the smooth AlN surface immediately followed by cooling under vacuum. (c) GaN quantum dots formed by depositing the equivalent of two GaN monolayers on the smooth AlN surface immediately followed by exposure to N plasma during 50 seconds. The structure reorganization or ripening effect is observed in (c). (Ref. 12)

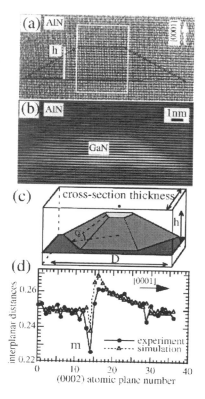

Fig. 4. (a) HRTEM image of a GaN dot, taken along [01-10] axis. The region of the truncated pyramidal shape of the dot is outlined. (b) Fourier filtered image of (a) obtained by using all the (000i) frequencies except the ones belonging to AlN. (c) Schematic view of the dot within the cross-section sample as deduced from fits: only half of the pyramidal dot occupies partly the cross-section thickness. (d) Experimental and simulated interplanar distance profiles of the dot. (Ref. 22)

particular case, as compared to the sample without 50 seconds ripening process, the dot density reduced from 5×10^{11} to 5×10^{10} cm^{-2}, while the average size (height/diameter) increases from 2/20 to 5/25 nm.

A detailed investigation with the help of reflection high energy electron diffraction (RHEED) shows a more significant ripening effect when the samples are exposed under vacuum than to nitrogen at growth temperature.[14] At lower temperature, however, the dot size and density remain unchanged under the nitrogen plasma. The reason was suggested that the Ga diffusion on the surface might be inhibited in the presence of nitrogen.[14]

The detailed structure of GaN QDs has been investigated by high-resolution transmission electron microscopy (HRTEM).[17,22] Figure 4a gives an example of the HRTEM image of a GaN dot embedded in AlN, taken along the [01-10] axis. A 3D schematic view derived from the HRTEM and RHEED analysis is shown in Figure 4c. The sample is composed of repeated layers of about 2 GaN MLs and thick AlN layers. The analysis of the HRTEM image reveals the following results: (1) The QD has a shape of truncated pyramid of hexagonal base with $\alpha = 30°$. The dimensions in this particular case are measured to be 3.3 nm in height and 15.3 nm in diameter. (2) The QD is fully

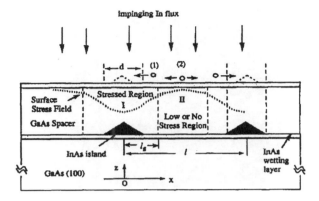

Fig. 5. A schematic representation showing the two major processes for the In adatom migration on the stressed surface: (1) directional diffusion under mechanochemical potential gradient contributing towards vertical self-organization and (2) largely symmetric thermal migration in regions from the islands contribution to initiation of new islands not vertically aligned with islands below. (Ref. 24)

strained and is dislocation free. (3) A wetting layer of 2 ML GaN is derived. Only a small amount of Ga and Al atoms (15% in this case) is exchanged across the GaN/AlN interface.

It has been established earlier that the strained islands such as InAs grown in successive layers separated by spacer layer such as GaAs could lead to vertical correlation if the thickness of the spacer layer is appropriate.[24,25] The driving force for such vertical self-organization is schematically shown in Figure 5.[24] Islands in the first layer produce a tensile strain in the spacer above the islands, whereas little stress exists in the spacer away from the islands. Indium adatoms impinging in the surface would be driven by the strained field to accumulate on the top of the islands where they can achieve an energetically lower state due to the lower lattice mismatch between the new islands and spacer. The vertical correlation of self-assembled multilayer QDs was also demonstrated in GaN/AlN system.[14,15] An example of HRTEM image is shown in Figure 6 (next page).[14] It reveals such a correlation for an AlN spacer of 8 nm. For a thicker AlN layer of 20 nm, no vertical correlation is observed.[12]

In addition to the strain-induced vertical-correlation of GaN QD arrays, a correlation between the QD growth and the threading edge dislocations propagating in AlN has been noted.[16] The conventional TEM and HRTEM images shown in Figure 7 (next page) demonstrate that the GaN QDs may be more likely to form adjacent to the edge dislocations. In this experiment, the dislocation density in the thick AlN layer is $1.8 \times 10^{11} cm^{-2}$, comparable to the density of GaN QDs ($1.1 \times 10^{11} cm^{-2}$). The strain field at the opposite side of the dislocation favors the nucleation of GaN QDs where the AlN lattice is stretched and the mismatch to GaN is smaller.[16] If the dislocation density is high, as in this case, the vertical correlation of the QDs may be disturbed by the presence of dislocation line that is slightly inclined. Instead of following the vertical positions of the QDs in the previous layer, the QDs seem rather likely to follow the dislocation line. This effect may be not important if the dislocation density is much lower than the QD density.

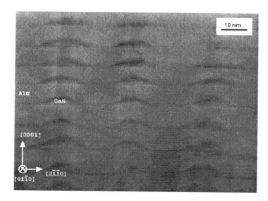

Fig. 6. HRTEM image, taken along the [01-10] direction of a superlattice of GaN dots capped by AlN. Because of the low magnification of the printed image, the atomic columns are not seen although they are present. The vertical correlation of the GaN dots is evident. The two-dimensional GaN wetting layer is also clearly visible. Note the dislocation line running through the column of dots at the left-hand side. (Ref. 14)

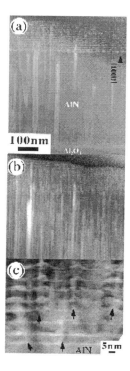

Fig. 7. (a) Weak-beam image of a cross section of the GaN QDs in AlN matrix (g={2,-1,-1,0}). Only dislocation with a Burgers vector along c =[0001] are visible. Most of them are screw dislocations. (b) Weak-beam image with g=(0001) of the same area. Only dislocations with an in-plane component Burgers vector of the form 1/3<2,-1,-1,0> are visible. (c) Slightly off-axis <0,1,-1,0> HRTEM image. The first eight QDs layer of the samples are seen. One can note that the shape of the QDs of the first three QDs layers above the AlN thick layer are less well defined than the shape of the QDs of the other layers. Traces of edge dislocation are outlined by dark arrows. The QDs are vertically correlated. (Ref. 16)

The growth of GaN QDs on Si(111)[18,26] and SiC(001)[27] substrates by MBE were also reported. The purpose of growing GaN QDs on Si substrates is mainly for the integration of light emitting devices with Si technology. The growth processes in the former case are essentially the same with those on sapphire substrates. By controlling the size of the GaN QDs in AlN matrix, intense room temperature PL with different colors from blue to orange as well as white were demonstrated (see next section).[18] When GaN was grown on 3C-SiC(001) surface, the QDs with cubic rather than hexagonal structure can be obtained. The zincblende GaN islands were formed on AlN buffer by rf MBE with an average height of 1.6 nm and a diameter of 13nm. The island density is $1.3 \times 10^{11} cm^{-2}$.[27]

In addition to SK mode, the 3D growth of GaN QDs on $Al_xGa_{1-x}N$ was possible by using a so-called the "anti-surfactant" Si.[19] In this experiment, a smooth $Al_xGa_{1-x}N$ layer was prepared on 6H-SiC(0001) by MOVPE and used as the substrate for MBE re-growth. The GaN QDs were grown by MBE in which NH_3 was used as N source and CH_3SiH_3 was used as a Si source. The $Al_xGa_{1-x}N$ surface was exposed to Si flux before the GaN growth, and the NH_3 flow was stopped for this step. The subsequent GaN growth was carried out with and without introducing Si on two different samples. In the sample without Si flux during the GaN growth, the growth was two-dimensional and the streaky RHEED patterns were observed. With Si flow, a change of GaN growth mode from 2D to 3D was observed and the RHEED patterns turned out to be spotty. Formation of GaN QDs was confirmed by AFM. The dot density could be changed by the variation of the Si flux and the growth temperature. The dot density decreased by a factor of 10^3 and the dot sizes increased from 4/50 to 10/200 nm by raising temperature from 660 to 740 °C.[19] More investigations on the anti-surfactant growth scheme by MOCVD will be presented in the next subsection.

For the applications to light emitting devices, the $In_xGa_{1-x}N$ alloy is more frequently used as the active layer. The band-gap and the emitting wavelength are easily modified by alloy composition. The quantum efficiency of light emission from $In_xGa_{1-x}N$ QWs is usually higher than GaN. As compared to GaN QDs, however, fewer investigations has been published on the growth of $In_xGa_{1-x}N$ QDs. The fluctuation in alloy composition or phase segregation during the growth may complicate the growth and the origin of light emission.

Recently, using the conventional MBE with rf plasma source, the growth of $In_xGa_{1-x}N$ QDs on GaN in SK mode has been demonstrated.[28] In this experiment, the substrate was a ~2μm thick GaN layer grown by MOCVD on sapphire. The growth parameters were monitored by RHEED and the In mole fraction was determined with an error of 3%. At the substrate temperature of 580°C, a layer by layer growth of $In_{0.35}Ga_{0.65}N$ on GaN was observed during the first 1.7 ML growth. Beyond 1.7 ML, the growth mode was changed from 2D to 3D and the $In_xGa_{1-x}N$ islands were formed. The AFM image of the $In_xGa_{1-x}N$ surface with 5 ML deposition shows a high island density of $\sim 10^{11} cm^{-2}$. The average diameter and height of the islands are 27 and 2.9nm respectively. This investigation shows that the transition of 2D-3D growth can be realized for In content from 18% to 100%. For an In content below 18%, the growth mode remains 2D. The multiple layers of $In_xGa_{1-x}N$ QDs can also be formed by the overgrowth of a GaN layer of 4-5nm which smooth the surface.

The $In_xGa_{1-x}N$ QDs grown by SK mode can also be realized by MBE using NH_3 as the nitrogen source.[29] Before the growth of $In_xGa_{1-x}N$, a GaN buffer of a few micrometers was grown at 820°C. $In_xGa_{1-x}N$ was then grown at temperatures 530-570°C and a growth rate of 0.1-0.2 μm/h. The In composition was kept at 0.15 which is larger than the critical

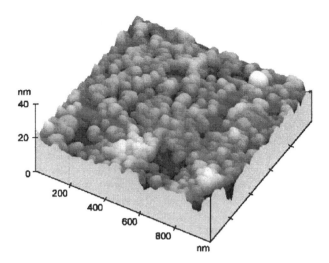

Fig. 8. 1μm×1μm AFM image of unburied self-assembled $In_xGa_{1-x}N$ islands grown on GaN by MBE. (Ref. 29)

value of 2D-3D transition (0.12) determined in the same experiments.[30] The 2D to 3D transition was observed after the deposition of 4-5 ML (~11Å) of $In_{0.15}Ga_{0.85}N$. The average island size is about 35 nm in diameter and 4nm in height. The island density was approximately $5\times10^{10} cm^{-3}$ and greater than the dislocation density in GaN buffer layer which was approximately $5\times10^{9} cm^{-3}$.

A typical surface morphology of $In_xGa_{1-x}N$ QDs on GaN imaged by AFM is shown in Figure 8. As compared to GaN, the density, and the sizes of $In_xGa_{1-x}N$ QDs are more difficult to control in growth. Except for AFM and RHEED investigations, no detailed lattice structures have been imaged by HRTEM for $In_xGa_{1-x}N$ QDs.

Instead of being active layers, GaN/AlN QDs can also be used as part of buffer layer to improve epilayer quality. Recently, we have investigated the GaN films grown on buffer layers containing QDs by MBE on sapphire substrates.[31] The density of the dislocations in the films was determined by defect delineation chemical etch and AFM.[32] It was found that the insertion of a set of multiple GaN QD layers in the buffer layer effectively reduces the density of the dislocations in the epitaxial layers. Figures 9 (a) and (b) (next page), respectively, show the AFM images of stained surfaces of a standard GaN and a GaN grown on quantum dot defect filter. The dislocation density in standard GaN on AlN buffer layer is on the order of 10^{10} cm^{-2} or higher which is typical. By contrast, the density in GaN on quantum dot filters is in the low 10^7 cm^{-2}. TEM observations showed disruption of the threading dislocations by the QD layers, although the microscopic nature of dislocation-QD interaction is not well understood.

2.2 MOCVD

The III-sources used in the MOCVD growth of III-nitrides are trimethyl-gallium (TMG), trimethyl-aluminum (TMA), and trimethyl-indium (TMI), carried by nitrogen (N_2) or hydrogen gas (H_2). Nitrogen source is from ammonia (NH_3). The self-organized growth of GaN QDs by MOCVD was first reported by Dmitriev et al[33] who grew GaN QDs

 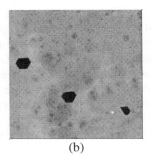

(a) (b)

Figure 9. AFM images of stained surfaces of a standard GaN (a - 1x1 μm² in size) and quantum dot filtered GaN (b - 2x2 μm² in size). The defect densities are ~ 10^{10} cm^{-2} for (a) and low 10^7 cm^{-2} for (b).

directly on 6H-SiC substrates. In this case, the lattice mismatch between the GaN and SiC is large enough to lead to the island growth. The main contributions of the growth of GaN QDs on $Al_xGa_{1-x}N$ (x<0.2) by MOCVD are by Tanaka and coworkers.[34,35,36,37] They developed a method called as the anti-surfactant which can change the growth mode from 2D to 3D. The self-assembly of GaN QDs is realized in this small lattice-mismatched system by exposing $Al_xGa_{1-x}N$ surface to Si during the growth. The Si is from tetraethyl-silane [$Si(C_2H_5)_4$:TESi:0.041 μmol] (TESi) and carried by H_2.

Samples were grown on Si face of 6H-SiC (0001) substrates.[34,35,36,37] Typically, after depositing a thin (~1.5 nm) AlN buffer layer, a thick (~0.6 μm) $Al_xGa_{1-x}N$ cladding layer was grown. The Al content x varied from 0.07 to 0.2. GaN was then grown on the top of this $Al_xGa_{1-x}N$ with a short supply (5 sec) of TMG and NH_3 during which TESi may or may not be used. The QDs may be covered by a ~60 nm $Al_xGa_{1-x}N$ layer for optical studies or left without capping layer for AFM image.[34] As shown in Figure 10, if TESi is

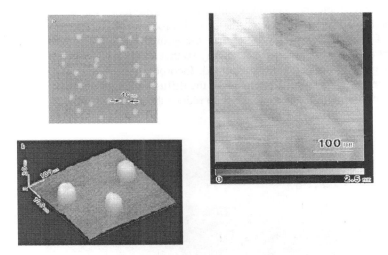

Fig. 10. AFM images of GaN quantum dots assembled on an $Al_xGa_{1-x}N$ surface using TESi as an anti-surfactant, (a) plane view; (b) bird view. (c) An AFM image of GaN grown on $Al_xGa_{1-x}N$ surface without TESi doping, showing a step flow growth. (Ref. 34)

not supplied, a step flow growth of GaN with a smooth surface was observed. Only when the $Al_xGa_{1-x}N$ surface is exposed to TESi, the GaN QDs were effectively grown. The step-flow growth observed without TESi flux (Fig. 10c) was explained by a fairly small lattice mismatch between GaN and $Al_xGa_{1-x}N$ (0.37 % for $x = 0.15$). Under the exposure of small Si dose, large GaN islands were formed. These islands transformed into isolated small dots under a higher Si flux (Figure 10a and 10b).

The dot density in this case could be controlled from ~10^7 to $10^{11} cm^{-2}$ by changing TESi doping rate, growth temperature, growth time, and alloy composition. It is very sensitive to the grow temperature, varying by a factor of 10^3 between 1060 and 1100°C. The dot size can also be controlled by a proper choice of growth parameters. Figure 11 gives the HRTEM image from typical GaN QDs grown on $Al_xGa_{1-x}N$ surface (x=0.2).[37] The hexagonal shaped GaN dots have an average height of ~ 6 nm and diameter of ~ 40 nm. The dot thickness/diameter ratio could be changed from 1/6 to 1/2 by varying the growth temperature and Si dose.[34,35] The dot density is ~$3 \times 10^9 cm^{-2}$, more than one order of magnitude lower than the GaN QDs grown by MBE. For a fixed growth temperature ($T_s = 1080$ °C) the densities of ~5×10^9 and ~5×10^8 cm^{-2} were obtained with TESi doping rate of 44 and 176 nmol/min, respectively. By increasing the GaN growth time from 5 to 50 seconds, the dot size was changed from ~6/40 to ~100/120 nm.[34] The PL and the stimulated emission[38] from the similar QDs will be discussed later.

Using a similar method, Hirayama et al.[39] fabricated $In_xGa_{1-x}N$ (x from ~0.22 to ~0.52) QDs on an $Al_xGa_{1-x}N$ (x=0.12) surface. A two-layer buffer structure was used. First a 300 nm $Al_{0.24}Ga_{0.76}N$ and then a 100 nm $Al_{0.12}Ga_{0.88}N$ layer was grown on a SiC substrate, both at 1100°C. Prior to $In_xGa_{1-x}N$ growth, a small amount of Si anti-surfactant was deposited at 1120 °C. Then the temperature was cooled to 800°C for QD growth. The dot density was as high as ~$10^{11} cm^{-2}$ and decreased with increasing Si dose. The average dot height and diameter were ~5 and 10 nm respectively, as determined from AFM images.

The microscopic mechanism of anti-surfactant in the growth of GaN QDs is not well understood. Incorporation of Si in the growth process is assumed to change the surface energy of the $Al_xGa_{1-x}N$ layer so that the growth mode is modified. An effective surfactant usually raises the surface energy so that a 2D layer-by-layer growth is favored. For GaN growth, the opposite is assumed. Incorporation of Si is assumed to reduce the surface energy of $Al_xGa_{1-x}N$ and increase the diffusion length of the adatoms. As a result, the adatoms are more likely to coalesce to reduce the total energy.

Fig. 11. Cross-section HRTEM image of uncapped GaN quantum dots grown on an $Al_xGa_{1-x}N$ surface. The upper bright part is glue used in the HRTEM sample preparation. (Ref. 37)

$In_xGa_{1-x}N$ QDs can also be grown on GaN by MOCVD without using anti-surfactant.[40] The equipment is an atmospheric-pressure two-flow system with horizontal quartz reactor. A GaN buffer layer was first grown on (0001)-sapphire at a temperature of 1075°C and a V/III ratio of 2000. The $In_xGa_{1-x}N$ portion was grown at a reduced temperature of 700°C with a growth time in the order of ~10s. The growth rate is estimated to be 0.17nm/s. The thickness of $In_xGa_{1-x}N$ is about 10 MLs. The average In composition is low (x<0.1). AFM images show the density of the QDs increasing with growth time. If the growth time is short (6.4ML), two kinds of QDs, bigger and smaller sizes were observed. The bigger QDs have a diameter of 15.5nm and height of 5.4nm. The smaller QDs have a diameter of 9.3nm and height of 4.2nm. When the growth temperature increases, the dot density decreases monotonically. The important difference in this case from the typical SK mode is the formation of the $In_xGa_{1-x}N$ QDs even at a long growth time (19ML). Thus, the formation of the QDs may be mainly attributed to phase segregation rather than strain-induced coalescence.

Formation of QD-like structures in semiconductor alloys and QWs induced by alloy fluctuation or phase segregation has a great effect on the material properties.[41] The In-rich clusters in $In_xGa_{1-x}N$ QWs were suggested to be the origin of high luminescence efficiency in $In_xGa_{1-x}N$/GaN LED.[42] Existence of alloy fluctuation/phase-segregation in $In_xGa_{1-x}N$ grown on GaN by MOCVD was confirmed by HRTEM.[41] The spherical QDs were observed in HRTEM images of a 280nm thick $In_{0.22}Ga_{0.78}N$ layer. A typical dot consists of a core and a surrounding strain zone. The lattice parameters inside the core are slightly larger than that in the surrounding matrix and approach those for InN. The size estimated from an HRTEM image is in the range of 1.5-3nm.

A method very different from self-assembly is selective growth of QDs. $In_xGa_{1-x}N$ QDs have been fabricated on Si-patterned GaN/sapphire substrates by MOCVD.[43,44] As shown in Figure 12 (next page), a Si film of 50nm was first deposited onto the surface of GaN/sapphire substrate by electron beam evaporation at room temperature. Nanoscale circular windows were then opened in the Si mask by focused ion-beam irradiation followed by photoassisted wet chemical etching. GaN/$In_xGa_{1-x}N$ multilayers were finally epitaxially grown on the GaN plinths with shapes of hexagonal pyramids. Both $In_xGa_{1-x}N$ QWs and QDs were formed in the structure but the QDs appeared only on the top of the pyramids. As compared to self-organized growth, the selective growth on patterned substrates could in principle provide a better way of controlling the position, size, and density of the QDs.

2.3 Other Techniques

In addition to MBE and MOCVD, growth of GaN QDs by other techniques was also reported. Goodwin et al.[45] have fabricated nanocrystalline GaN by reactive laser ablation of pure Ga metal in a high purity N_2 atmosphere. The samples were collected from the surface of a membrane filter and then thermally annealed at 800°C in a high-purity ammonia atmosphere. TEM dark field images show a log-normal size distributions with a mean diameter of 12 nm and a standard deviation of 8 nm. Selected-area electron diffraction pattern confirms the hexagonal phase. The quantum confinement effect was observed from the blue shift of the size-selective PL and PLE spectra. Nanocrystalline GaN thin films were also fabricated recently on quartz substrates by rf sputtering using GaAs as a target material at a nitrogen pressure of 3.5×10^{-5} bar.[46] The average particle

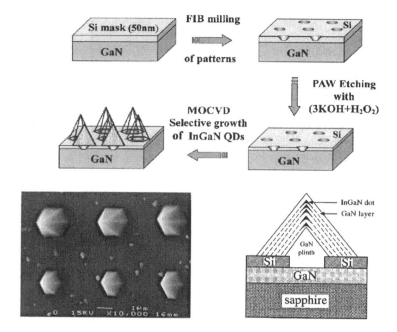

Fig. 12. Upper panel: Schematic diagram of the focused-ion-beam/MOCVD process used for the fabrication of $In_xGa_{1-x}N/GaN$ nanostructures. 60% of the Si layer was sputtered by focused-ion-beam whereas the rest was removed by photo-assisted-wet etching. Five periods of $In_xGa_{1-x}N/GaN$ QDs were then selectively grown on GaN plinths exhibiting a small density of dislocations. Lower left panel: A top view of SEM image of GaN plinths laterally overgrown on circular windows with diameters of 600 nm (upper row) and 300 nm (lower row). The regrown GaN has a hexagonal pyramid shape with six {1,-1,0,1} side facets. Lower right panel: Schematic drawing of $In_xGa_{1-x}N$ quantum dot structures. (Refs. 43 and 44)

size of the nanocrystalline GaN increased from 3 to 16nm when the substrate temperature was raised from 400 to 550°C.

Crystalline GaN particles can be synthesized by simple inorganic reactions at various temperatures. Well et al.[47,48] reported a method of nanosized GaN synthesis by pyrolysis of gallium imide $\{Ga(NH)_{3/2}\}_n$ at high temperatures. Dimeric amidogallium $[Ga_2N(CH_3)_2]_6$ was first synthesized by mixing anhydrous $GaCl_3$ with $LiN(CH_3)_2$ in hexane. This dimer was then used to prepare polymetric $\{Ga(NH)_{3/2}\}_n$ by reacting $[Ga_2N(CH_3)_2]_6$ with gaseous NH_3 at room temperature for 24 hours. The GaN QDs were prepared under ammonia flow by slowly heating $\{Ga(NH)_{3/2}\}_n$ in trioctylamine to 360°C, cooling it to 220°C to add and stir a mixture of trioctylamine and hexadecylamine, and finally cooling it to room temperature. Formation of isolated spherical QDs in colloidal GaN solution was confirmed by TEM images, shown in Figure 13 (next page).[49] The image reveals that the GaN dots have a zinc-blende structure with diameters ranging from 2.3 to 4.5 nm. The absorption and PL peak were observed to shift to a higher energy as compared to bulk GaN. In addition to GaN, an $Al_xGa_{1-x}N$ nanoparticle/polymer composite was also synthesized using a similar method and the microstructure of zinc-blende QDs was confirmed by HRTEM.[50]

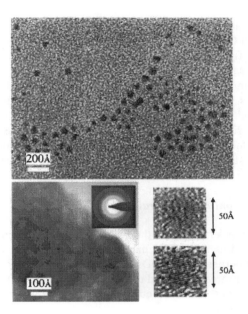

Fig. 13. TEM image of GaN QDs taken in bright field. The particles are well dispersed and not agglomerated. Top panel shows low magnification of QDs and some linear alignment. Bottom two right panels show high magnification and lattice fringes of QD oriented with the <111> axis in the plane of the micrograph. Bottom left panel shows electron diffraction pattern of GaN QDs indicating zinc-blende structure. (Ref. 49)

3. Optical Properties of III-Nitride QDs

Optical properties of III-Nitride QDs were mostly obtained by photoluminescence (PL). Few other investigations have been reported up to date. The reported PL spectra show very different characteristics for the different samples prepared by different methods. For the GaN QDs alone, the published PL peak energies may be much lower or higher than the band gap of bulk GaN, ranging from 2.15 to 3.9eV depending on sample and measurement temperature. In Figure 14, we present the PL spectra from five different

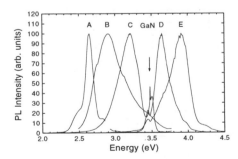

Fig. 14. PL spectra from five GaN/AlN QD samples grown by MBE on different substrates under different growth conditions. Sample A and B were grown on Si substrates while sample C, D, and E on sapphire substrates. The spectra were measured at ~15K under the excitation of a Ti-sapphire laser (photon energy 5.06 eV). Intensity is normalized at the maximum of the QD-related signal.

samples containing GaN/AlN QDs grown by MBE.[51,52] The weak peaks near 3.45eV are the band edge PL from GaN bulk. The strong broad peaks are attributed to the PL from QDs. As shown in the figure, the QD-related peaks have different energies from ~2.6 to 3.9eV for the different samples.

In Table 2, we summarize the PL peak energies from most of the published spectra dealing with GaN QDs. The measurement temperature, matrix and substrate materials, preparation method, and the size of QDs are also listed. This seemingly large discrepancy may in fact reflect the rich and interesting properties of III-nitride QDs. In this section, we will first describe three important factors pertinent to III-nitride QDs, the effects of quantum confinement, strain, and polarization. Then, we will discuss in detail the PL and other relevant properties of GaN and $In_xGa_{1-x}N$ QDs grown by various techniques.

3.1 *Effects of quantum confinement, strain, and polarization*

The quantum confinement effect shifts the band gap of a bulk semiconductor to higher energy. This shift, called as the confinement energy, depends on the size and shape as well as the material properties of both QDs and surrounding matrix. Here we estimate the confinement energies for three simplified cases: a plate or disk, a cubic box, and a sphere. Assuming an infinite barrier, the confinement energy of the ground state for an electron in a rectangular box is given by

Table 2. The reported peak energies of the photoluminescence from GaN quantum dots and nanocrystallines. The measurement temperature, matrix material, substrate, growth method, dot size (growth thickness, height/diameter ratio, or diameter), and references are also listed.

E_{PL} (eV)	T (K)	QDs	Matrix	Substrate	Growth Method	Size(nm) Height/Dia.	Ref.
2.15	300	GaN	AlN	Si(111)	MBE	12MLs[a]	18
2.50						10MLs	
2.80						7MLs	
2.95	2	GaN	AlN	Sapphire	MBE	4.1/17	15
3.75						2.3/8	
2.95	300	GaN	AlN	Sapphire	MBE	4/16	16
3.75	2, 300	GaN	AlN	Sapphire	MBE	4/20	14
3.47	80	GaN	$Al_{0.15}Ga_{0.85}N$	6H-SiC	MOCVD	40/120	37,68
3.50						7/21	
3.58						3.5/10	
3.45	340	GaN	$Al_{0.15}Ga_{0.85}N$	6H-SiC	MOCVD	7/21	36
3.53	77	GaN	$Al_{0.1}Ga_{0.9}N$	6H-SiC	MOCVD	1-2/10	38
3.55	77	GaN	$Al_{0.2}Ga_{0.8}N$	6H-SiC	MOCVD	6/40	34
3.47	80	GaN	$Al_{0.12}Ga_{0.88}N$	6H-SiC	MOCVD	40	69
3.8	10,300	GaN[c]	AlN	3C-SiC	MBE	1.6/13	27
3.8		GaN[n]			Ablation	12	45
3.4	15	GaN[n]		Quartz	Sputtering	16	46
3.5						10	
3.75						7	
3.9						3	

a: Grwoth thickness in units of monolayer (ML)
c: Cubic structure
n: Nanocrystalline

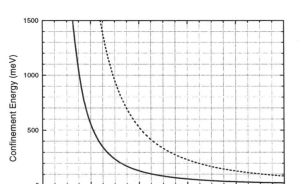

Fig. 15. Shift in the band-gap energy from bulk GaN for a 2D GaN plate of thickness d (solid line) and a GaN sphere with a diameter d (dotted line) as a function of d. The shift in the band-gap energy in a cubic box of size d is between these two curves.

$$E = \frac{h^2}{8}(\frac{1}{m_x d_x^2} + \frac{1}{m_y d_y^2} + \frac{1}{m_z d_z^2})$$

where d_j (j=x, y, and z) are the dimensions of the box and m_j are the electron mass in j directions. For a plate- or disk-like dot, in which the in-plane sizes d_x and d_y are much larger than the height $d_z = d$, the confinement energy is simplified to $h^2/(8m_z d^2)$. The shift in the band gap is calculated using $1/m_z = 1/m_{e,z} + 1/m_{h,z}$, where $m_{e,z}$ and $m_{h,z}$ are respectively the effective masses of the electrons and holes in III-Nitride along the z-direction. For a cubic box of size d, the confinement energy is given by the same expression, $h^2/(8m_z d^2)$, but with $1/m_z$ replaced by $(2/m_{e,t} + 2/m_{h,t} + 1/m_{e,z} + 1/m_{h,z})$, where $m_{e,t}$ and $m_{h,t}$ are the transverse effective masses of the electrons and holes. For a sphere of diameter d, the ground state energy is given by $h^2/(2m_z d^2)$ if an isotropic effective mass is assumed.[53] In Figure 15, we plot the confinement energies as a function of d for a plate and a sphere. In both cases, the effective masses of $0.22m_0$ for electrons and $1.1m_0$ for holes were assumed. The confinement energy in a cubic box will be between these two curves. As the results show, for small dots of a few nm in size, the confinement energy is very sensitive to the dot size. This sensitivity is much reduced when the size is increased. Quantitatively, when the size decreases from 10nm to 2nm, the confinement energy may change from 20meV to more than 1eV depending on the shape of the dots.

The curves shown in Figure 15 represent good lower and upper bounds for the real confinement energy in GaN QDs, both free and embedded in AlN matrix, if the size is not too small. For electrons, the barrier height in GaN/AlN interface (75%ΔE_g or 2.1 eV)[54] is really high and the mass anisotropy is small. For holes, the barrier is lower (0.7eV) and the mass anisotropy is larger. However, the effect on confinement energy is much reduced due to the large masses of holes. For a specific GaN QD, the actual energy shift from the bulk value is expected to be within those two bounds mentioned earlier. In the case of self-assembled QDs, the plate-like results may be more suitable. In other cases such as GaN nanocrystallines, sphere-like shape may be more appropriate.

The effect of strain in III-nitride heterostructures must be taken into account. For GaN QDs grown on $Al_xGa_{1-x}N$ and $In_xGa_{1-x}N$ QDs on GaN, the strain is commonly compressive which induces a blue shift in the band gap. This shift is proportional to the strain and the relevant deformation potentials. In the case of vanishing shear strain, the bandgap shift can be conveniently expressed as $\Delta Eg = (2\alpha+\beta q)\varepsilon_{xx}$ or $(2\alpha/q+\beta)\varepsilon_{zz}$,[55,56] where $q=-2C_{13}/C_{33}$, C_{13} and C_{33} are the elastic constants, ε_{xx} and ε_{zz} are the xx and zz components of the strain tensor, and α and β are the constants related to the various deformation potentials of both conduction and valence bands. The reported value of $(2\alpha+\beta q)$ is from -6 to -12 eV,[55,56,57,58] depending on measurements and the choice of q (from -0.51 to -0.60).[59] Assuming $(2\alpha+\beta q) = -10$ eV, we obtain $\Delta Eg = -10\varepsilon_{xx}$ eV. It is 0.25eV in the case of fully strained GaN on AlN. For a fully strained GaN on $Al_xGa_{1-x}N$, assuming a linear dependency of $Al_xGa_{1-x}N$ lattice constant on x, the energy shift $\Delta Eg = 0.25x$ eV.

As compared to the strain induced by lattice mismatch, the strain induced by thermal mismatch is negligible. For a rough estimation, we use a temperature difference of 1000 K between the growth and the measurements. Using the thermal expansion coefficient $\Delta a/a$ of 5.6×10^{-6} K^{-1} (see Table 1) for GaN and 7.5×10^{-6} K^{-1} for sapphire,[60] a compressive strain of about 0.002 is obtained. This strain will only produce a blue shift in the band-gap of 20 meV.

In contrast to the effect of confinement and strain, the polarization effect may induce a red shift in the band-gap. This effect is very important in the III-Nitride semiconductors and has been discussed extensively in literature.[54,61,62] Both spontaneous polarization and strain induced polarization (piezoelectric effect) produce large electric field which has significant effect on the optical properties of QDs. From the symmetry arguments, it can be concluded that the wurtzitic (hexagonal) crystal has the highest symmetry showing spontaneous polarization.[63] Bernardini et al.[64] have calculated the spontaneous polarization P_{Spon} and the piezoelectric constants e_{31} and e_{33} for III-Nitrides. By defining the spontaneous polarization as the difference of electronic polarization between the wurtzitic and zincblend structures, they obtained P_{Spon} from first-principle calculations. The results are listed in Table 1 along with the piezoelectric constants for GaN, AlN, and InN. As compared to GaAs, the piezoelectric constants of III-nitrides are about an order of magnitude larger.[64]

The electric field associated with the polarization can be equivalently described by the bulk and interface polarization charges, $\rho_P(r)=-\nabla\cdot P(r)$ and $\sigma_P=-(P_2-P_1)\cdot n_{21}$. Here n_{21} is the direction of the interface pointing from medium 1 to 2 and (P_2-P_1) is the difference of the polarization across the interface. Consider a thick GaN layer grown on AlN (0001) surface, the interface charge due to the difference in spontaneous polarization in two crystals is -0.052 C/m^2 corresponding to a carrier density of $3.3\times10^{13}cm^{-2}$. The electric field E_P created by this polarization charge is $\sigma_P/(2\varepsilon_r\varepsilon_0)$, here $\varepsilon_r\varepsilon_0$ is the static dielectric constant. It is -2.8×10^6 V/cm in GaN and 3.5×10^6 V/cm in AlN, with the sign reference to the growth direction. For a thin GaN plate straddled by AlN, as schematically shown in Figure 16 (next page), the polarization charges in both bottom and top interfaces must be considered and the result is doubled (-5.6×10^6V/cm).

Additional polarization induced by strain can be calculated from the elastic and piezoelectric constants listed in Table 1, $P_z=2e_{31}\varepsilon_{xx}+ e_{33}\varepsilon_{zz}=2(e_{31}-e_{33}C_{13}/C_{33})\varepsilon_{xx}$. For a fully strained GaN plate on bulk AlN, P_z is calculated to be 0.033 C/m^2. It should be

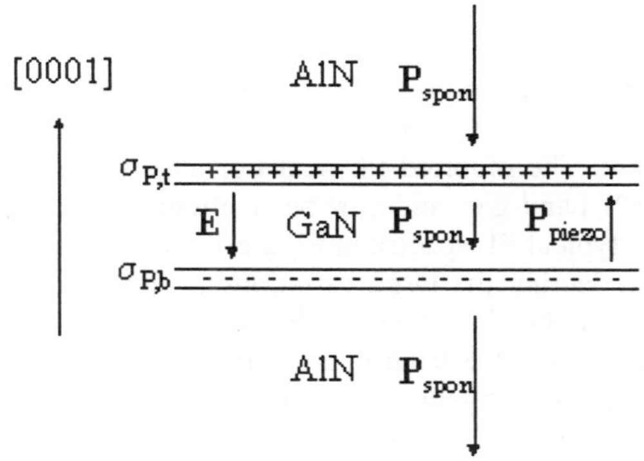

```
          [0001]                  AlN    P_spon ↓

                  σ_P,t  +++++++++++++++++++++++
                         E ↓   GaN   P_spon ↓   P_piezo ↑
                  σ_P,b  - - - - - - - - - - - - - - - - -

                                  AlN    P_spon ↓
```

Fig. 16. A schematic diagram of a GaN plate grown on and covered by thick (0001) AlN, showing the polarization effect. P_{spon} is the spontaneous polarization that is -0.029 C/m^2 in GaN and -0.081 C/m^2 in AlN. P_{piezo} is the polarization induced by strain (2.5% in in-plane direction in GaN), which is 0.033 C/m^2 only appearing in GaN layer. $\sigma_{P,b}$ and $\sigma_{P,t}$ are the polarization charges at the bottom and top interfaces, which are -0.085 C/m^2 and 0.085 C/m^2 respectively. E is the electric field which is -9.2×10^6 V/cm. Notice the directions of P and E with respect to [0001] and the signs of the interface charges.

reminded that the effect of spontaneous and strain induced polarization in GaN/AlN heterostructures is additive. As a result, the total electric field induced by spontaneous polarization and piezoelectricity is raised to -4.6×10^6 V/cm if only a single interface is considered. This field is doubled (-9.2×10^6 V/cm) in a thin GaN plate straddled by AlN (See Figure 16). This huge field will drive free electrons in the GaN toward the top interface and holes towards the bottom interface. As a result of the quantum confined stark effect (QCSE),[65,66] the electron-hole transition energy is greatly reduced.

The field-induced shift in transition energy in quantum wells has been investigated extensively by Miller et al.[66] The shift is more significant for a wider well than a narrower one, particularly if the linear approximation holds. In much wider wells, a quadratic approach which reduces the red shift is used. In addition, if the Fermi levels come close to the conduction band on one side and valence band on the other, the resultant free carriers would screen the induced field, reducing the shift. A similar result is expected for plate-like QDs. For a rough estimate, the shift is approximately E_Fd in the limit of large size. This gives a value of 2 eV for a plate of 4nm thick under a field of 5×10^6 V/cm.

In the following subsections, we will discuss various PL results encompassing the quantum confinement, strain, and polarization effects. Only when these effects are properly considered, can the experimental results be reasonably understood. To reiterate, the quantum confinement and strain effects essentially induce a blue shift, while the polarization effect results in a red shift. In addition, the quantum confinement is accentuated in small dots, while the polarization effect is more significant in large and strained dots. Experimentally, it is possible to distinguish various effects at least qualitatively. In the case of GaN QDs, if the observed PL has a lower energy than the bulk GaN bandgap, the polarization effect must be dominant. Otherwise, the quantum confinement effect is more important.

3.2 GaN quantum dots

The GaN QDs grown by laser ablation,[45] rf sputtering,[46] and chemical synthesis[49] are the examples in which the strain and polarization effect seem to be less important as compared to confinement. The reported PL peaks or absorption edges in all these cases are higher than the GaN band-gap and can be qualitatively attributed to the quantum confinement effect. The typical PL spectra in Figure 17 show broad peaks with a width of a few hundreds meV, reflecting the large variations in dot size. In addition, noticeable Stokes shifts of the PL from the absorption (or PLE) peaks are often observed.[45]

Preschilla et al.[46] estimated the band gap of GaN QDs with different sizes from the absorption spectra. In this investigation, the nanocrystallites were deposited on quartz substrates by reactive rf sputtering. The GaN particle size was found to change with substrate temperature and the structures were measured by XRD and TEM. The measured absorption edges were 3.4, 3.6, 3.9, and 4.1 eV for the average size of 16, 10, 7, and 3 nm respectively. These energies agree reasonably well with those obtained from the PL spectra. As compared to the curves in Figure15, the measured energy shifts from the bulk GaN band-gap are close to the confinement energies of a spherical dot, except for the 7 nm sample. The measured shift in the latter case (0.5eV) is much higher than that would be expected theoretically (~0.2eV).

The GaN QDs grown by MBE are more disk-like and their size and shape are more controllable. Daudin et al.[14,16,17] investigated the PL spectra, in more detail, of self-assembled GaN QDs embedded in AlN matrix grown by MBE using the SK mode. Two very different cases were observed, as shown in Figure 18. In the case of what is called "small dots" of 2.3 nm height and 8 nm diameter, a PL peak at 3.7 eV was observed at 2K. This is blue shifted from the bulk GaN band gap by 0.2 eV. A similar PL peak (~3.8eV) was also reported from the "small" GaN QDs (1.6 nm in height and 13 nm in diameter) with cubic structure grown by MBE on 3C-SiC(001) substrate.[27] However, in the case of what is called "large dots" of 4.1 nm height and 17 nm diameter, the detected PL peak is 2.95 eV, nearly 0.5 eV lower than the bulk band gap.[14,16,17] Considering the blue shift caused by quantum confinement (~0.1eV, see Figure 15) and strain (~0.2eV, see the discussions in the last subsection), the net red shift is even larger (~0.8eV). This

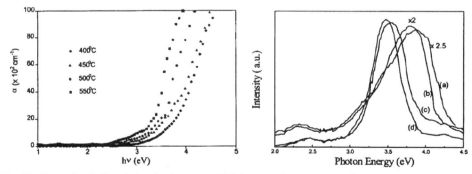

Fig. 17. Absorption (left) and photoluminescence (right) spectra from nanocrystalline GaN films deposited by rf sputtering at 400°C (a), 450°C (b), 500°C (c), and 550°C (d). The average particle size is found to correlate with growth temperature, which are 3, 7, 10, 16 nm at growth temperature of 400°C, 450°C, 500°C, and 550°C respectively. The band gaps derived from the absorption spectra are respectively 4.1, 3.9, 3.6, and 3.4 eV in the order of increasing particle size. (Ref. 46)

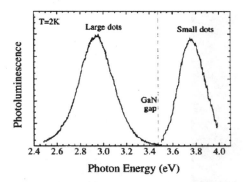

Fig. 18. Photoluminescence spectra of GaN quantum dots with a large (height/ diameter = 4.1/17 nm) and small (2.3/8 nm) average size, measured at 2K. Note the large red shift observed for the large dots with respect to small ones. (Ref. 15)

shift was interpreted as the result of the electric field induced by the polarization. As was discussed earlier, this field could be on the order of 10 MV/cm and may induce a large Stark shift in transition energy. A simplified calculation using a 2D QW of thickness H=0.72h (h being the height of the QDs) shows that the 5.5MV/cm electric field can account for the measured PL peaks from both large and small dots.[15]

A more detailed calculation reported recently confirms the above results.[67] In this calculation, a realistic geometry of the QDs measured from TEM was used. The effects of strain, quantum confinement, and polarization were all considered. In addition to the ground state energy, the distributions of strain, electric field, and charges in the QDs were also obtained. Figure 19 presents the electron and hole distributions in the QD for the ground states and the first excited states.[67] As described before, the electrons are more localized near the top interface, while the holes are near the bottom interface. It should be emphasized that the carrier distribution shown in Figure 19 is only correct for GaN grown

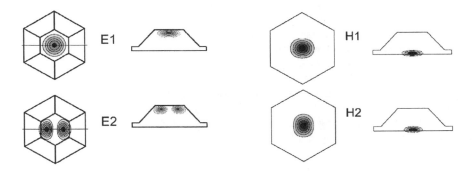

Fig. 19. Probability density distribution for the two lowest electron and hole states in a GaN quantum dot embedded in (0001) AlN matrix, calculated in the framework of an 8-band k·P model with zero spin-orbit splitting. The GaN dot has a similar shape as displayed in Fig. 4c. Note that the electron is mainly localized near the top interface, while the hole is near the bottom interface. (Ref. 67)

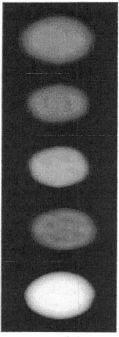

Fig. 20. Photographs of the light emitted at room temperature from GaN/AlN QDs on Si(111) substrate excited by a 10 mW unfocused He-Cd laser (~0.3 W/cm^2). (Ref. 18)

— 1 mm

on (0001) AlN face, called Ga-face (see Figure 16). If the GaN QDs are grown on (000-1) AlN face, called N-face,[21] then the electrons are expected to be driven towards the bottom interface and the holes towards the top interface.

The polarization effect is experimentally supported by the dependence of the PL spectra on excitation density. When the excitation density increases, the photo-carriers can partially screen the polarization field. As a result, the built-in electric field decreases and the transition energy increases as observed experimentally.[14,51]

The polarization effect was also used to interpret the origin of the blue, green, and orange PL (observed at room temperature, see Figure 20) from the GaN QDs in AlN matrix grown on Si(111) substrates.[18] These QDs were formed by growth of 7, 10, and 12 GaN MLs on AlN. The actual dot sizes were not reported. The PL peaks corresponding to different sizes were measured at about 2.8, 2.5, and 2.2 eV, all well below the band-gap of bulk GaN. It is interesting to note that, in the largest QDs of 12 MLs, the red shift is as large as 1.2 eV. By growing multilayers of GaN QDs with different sizes, a white light emission is demonstrated.

Here, we should emphasize that, in real QDs, the polarization effect is also expected to be very sensitive to the shape, doping, and interface properties of the materials, since the polarization field may be screened by any of the other charges associated with free carriers and defects. These complications may account for some apparent discrepancies reported in the literature. For example, higher PL and CL peaks (~3.8eV) than the GaN band-gap are reported from the GaN QDs in AlN matrix grown by MBE using the SK growth mode.[14] The size and the shape of the QDs measured from HRTEM, about 4 nm in height and 20 nm in diameter, are similar to the "large" dots mentioned above.

Although the sample had multilayer structures, the vertical correlation was negligible since the AlN spacing layer was sufficiently thick.

The PL spectra reported by Tanaka et al. associated with the GaN QDs in $Al_xGa_{1-x}N$ matrix are all near or higher than the GaN band-gap, ranging from 3.45 to 3.58 eV depending on the dot size and measurement temperature.[34,36,37,38,68,69] The QDs were grown on $Al_xGa_{1-x}N$ with x ~ 0.15 by MOCVD using Si as "anti-surfactant". The dot shapes are all disk-like and the sizes are mostly large, varying from 3.5 to 40 nm in height and 10 to 120 nm in diameter.[34,36,37,68] A stimulated emission peak at 3.49 eV was also reported at 20 K from small dots of 1~2 nm in height and 10 nm in diameter.[38] For large QDs, the energy shift due to quantum confinement may be less than 0.2 eV (see Figure 15). The shifts due to the strain and the strain-induced polarization are negligible since no strain is expected in these QDs grown through the use of an anti-surfactant.

Although the polarization effect is not discussed in these papers, it can be estimated assuming a linear dependency of the spontaneous polarization on x in $Al_xGa_{1-x}N$. We should note that this linear dependence is only an assumption for the sake of discussion. For x=0.15, following the discussion in the last subsection, a polarization charge of 0.017 C/m^2 or a carrier density of $1.0 \times 10^{13} cm^{-2}$ is calculated for the bottom interface (GaN on (0001) $Al_{0.15}Ga_{0.85}N$ interface). The same amount of negative charge is expected at the top interface. The electric field in this case is ~1MV/cm if the polarization charge from one interface only is considered. This value is doubled if both bottom and top interfaces are considered. The result will be different if the GaN QDs are covered by AlN instead of an alloy. As compared to the fully strained GaN QDs in AlN matrix grown using the SK mode, the polarization-induced electric field in this case is about 5 times smaller. The actual field may be lower due to Si doping, which is not well understood. As a result, the Stark shift due to this polarization may be small. It is difficult to get a more quantitative comparison. Any measured transition energy for a particular assembly of QDs must be the result of competition between the blue shift induced by quantum confinement and the red shift induced by polarization. It is expected to be not very far from the bulk GaN band-gap, as observed in the PL spectra.

The width of the PL spectra from GaN QDs ranges from a few tens to a few hundreds meV, depending on sample and measurement temperature. As in the case of the peak energy, the published PL widths and their variation cannot be explained by quantum confinement alone. If only the quantum confinement is considered, a QDs assembly with large average size would give narrow PL peak, as observed in Refs 36, 37, and 46 (see Figure 17 for example). However, when the polarization effect becomes important, the opposite holds. The QDs with a larger average size give a wider PL spectra (see, for example, Figure 18).[15,18] In the latter case, the energy fluctuation of different QDs due to different Stark shift has a more significant contribution to the PL line-width. It is interesting to note that a CL of ~50meV width was reported from a single GaN QD at 80K, which is much broader than expected.[69] Although the origin of this widening is not clear, the result suggests that some other effect such as background doping may also have an important contribution to the CL (likewise PL) width.

One of the most important properties distinguishing QDs from QWs and bulk materials is the better thermal stability of PL. This property is clearly demonstrated in the GaN QDs embedded in AlN matrix. Typically, the PL intensity from bulk materials quenches rapidly as the temperature is raised from low to room temperature. In the case of GaN QDs, a less temperature dependency is commonly observed. An example is given by Widmann et al. who compared the PL from GaN QDs grown on GaN substrate at 2

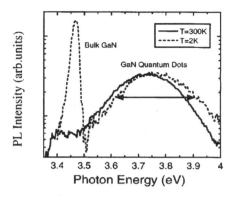

Fig. 21. Photoluminescence spectra from GaN QDs embedded in AlN matrix grown on GaN substrate, measured at 2 and 300K, showing the higher thermal stability of the PL from QDs than that from GaN substrate. (Ref. 14)

and 300 K.[14] As shown in Figure 21, the PL spectra at 2K are dominated by two peaks at 3.46 and 3.75 eV, from GaN substrate and QDs respectively. When the temperature was raised to 300K, the PL from GaN substrate totally disappeared, while the intensity of the PL from the QDs remained the same. A temperature dependency of the PL spectra from the cubic GaN QDs in AlN matrix was reported by Martinez-Guerrero et al.[27] When the sample temperature was raised from 10 to 200K, the integrated PL intensity changed very little. When the temperature was further increased to 300K, an increase in the PL intensity rather than decrease was observed. The latter phenomenon was attributed to the increase in the laser absorption resulting from a reduction in GaN band-gap with temperature.

Thermal stability of the GaN QDs in $Al_xGa_{1-x}N$ matrix is not as good as that in AlN.[36,37,68] Typically, when the temperature is raised from 10 to ~300K, a decrease in the PL intensity from a few to more than 10 times is observed. Although the PL intensity and its change with temperature may also depend on the quality of the material, the poorer thermal stability in the $GaN/Al_xGa_{1-x}N$ QDs compared to the GaN/AlN QDs can be interpreted by the lower barrier and the poorer confinement of the carriers.

The binding energy of A excitons in bulk wurtzite GaN was measured to be 21 meV.[70,71] In quantum dots, the electrons and holes must be closer to each other due to confinement. Thus, the electron-hole Coulomb interaction as well as the exciton binding energy should increase. From the temperature dependency of the PL peaks observed in $GaN/Al_xGa_{1-x}N$ QDs with different sizes, an increase in the exciton binding energy as a function of QD size was derived by Ramvall et al.[37,68] The results are shown in Figure 22. As the QD height is reduced from 7.1 to 1.25 a_x (a_x=2.8nm being the Bohr radius of the excitons in bulk GaN), or from 20 to 3.5 nm, an increase in the exciton binding energy as large as ~20meV is found. The experimental results are supported by the variational calculation of the excitons in two-dimensional structures.

In addition to the exciton binding energy, Ramvall et al.[36,37] also investigated the electron-phonon coupling in $GaN/Al_xGa_{1-x}N$ QDs. By fitting the measured temperature dependency of the PL peaks using the theoretical expression, $\Delta E_g = -\alpha_0[f(\hbar\omega_0)+1]$, they derived the electron-phonon coupling constant α_{0QD} relative to bulk value for the QDs

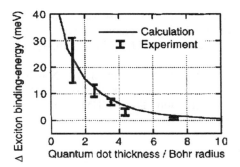

Fig. 22. The binding energy of the excitons in GaN QDs in $Al_xGa_{1-x}N$ matrix as a function of dot size, showing an increase in the binding energy with reducing dot size. (Ref. 37)

with different sizes. In the above expression, $f(\hbar\omega_{LO})=1/[\exp(\hbar\omega_{LO}/kT)-1]$ is the distribution function of the LO phonons and $\hbar\omega_{LO}=91.7$ meV. They found that $\alpha_{0QD}/\alpha_{0Bulk}$ is close to 1 for the QDs of size (height/diameter) of 12/36 nm. This ratio decreases to 0.81 and 0.63 as the size is reduced to 10/30 and 7/21 nm respectively. A decrease of electron-phonon coupling for small QDs is expected when the separation of the discrete energy levels is large and very different from the phonon energies.

Except for light emission properties, very few other optical spectroscopy has been used to investigate GaN QDs. Recently, Gleize et al.[26] reported the Raman spectra from two different samples of multilayer (80 and 88 periods) GaN QDs embedded in AlN matrix, grown by MBE using SK mode. Under the excitation in the visible range, far from the resonant condition, they observed new peaks at 606 cm^{-1} from sample A and 603 cm^{-1} from sample B (see Figure 23 for the spectra from sample B). Both peaks were assigned to the E_2 phonons in GaN QDs rather than any disorder activated scattering. This assignment was arrived at from the selection rules and a comparison to the Raman

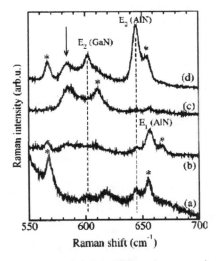

Fig. 23. Micro-Raman spectra of GaN quantum dots in AlN matrix measured at room temperature using 647 nm laser line, in four different configurations. (a) z(xy)-z, (b) x(yz)-x, (c) x(zz)-x, and (d) x(yy)-x. Asterisks mark phonons from the buffer layers. Features at about 618 cm^{-1} are related to the Si(111) substrate. (Ref. 26)

spectra from an $Al_xGa_{1-x}N$ film. From the observed peak positions, the biaxial strain in the QDs was estimated to be −2.6% for sample A and −2.4% for sample B. The results are consistent with the fully strained GaN QDs in AlN matrix as measured by HRTEM in similar samples.

The stimulated emission from GaN QDs in $Al_xGa_{1-x}N$ matrix was demonstrated at 21 K with optical pumping.[38] An emission peak was observed at 3.48 eV, about 50meV lower than the spontaneous emission. The FWHM was ~10meV. The threshold pump density was 0.75 MW/cm^2.

3.3 InGaN quantum dots

As compared to GaN, much fewer optical studies have been reported on $In_xGa_{1-x}N$ QDs grown on GaN by MBE or MOCVD. Instead, more investigations were focused on the origin of the PL from $In_xGa_{1-x}N$ QWs in which the In-rich structures can be formed through phase segregation or alloy fluctuation. It is believed that the intense PL from $In_xGa_{1-x}N$ wells may be associated with dot-like structures. In this subsection, in addition to the PL properties of $In_xGa_{1-x}N$ QDs, we will also discuss some relevant features of $In_xGa_{1-x}N$ QWs.

The reported PL peaks from $In_xGa_{1-x}N$ QDs range from 2.2 eV to a value higher than 3 eV, depending on the sample and measurement temperature.[29,39,40,43,44,72] All three effects discussed in GaN QDs, the quantum confinement, strain, and polarization, are also important in $In_xGa_{1-x}N$ QDs. The differences of the two systems are mainly rooted in the alloy nature of the latter material. The band-gap of $In_xGa_{1-x}N$ is sensitive to the alloy composition. A 1% fluctuation in x will lead to a shift in band-gap by 15meV if a linear dependency of $E_g(In_xGa_{1-x}N)$ on x is assumed. For x~10%, this shift can be much larger (~40meV) due to the large bowing factor of the alloy.[73] Since it is hard to determine the accurate value of x and its spatial fluctuation, an uncertainty in transition energy is introduced, making it difficult to derive reliable information from the energies of the measured PL peaks that are also affected by confinement, strain, and polarization effects.

Typical PL spectra from $In_xGa_{1-x}N$ QDs in GaN matrix were reported by Damilano et al. (Figure 24).[29] The self-assembled QDs were grown on GaN by MBE using the SK

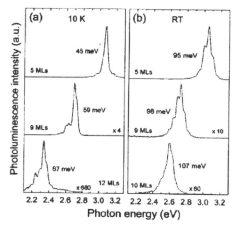

Fig. 24. (a) 10K and (b) room-temperature PL spectra of $In_xGa_{1-x}N$ QDs in GaN matrix of varying sizes. The nominal $In_xGa_{1-x}N$ thickness is 5, 9, 12 MLs in (a), and 5, 9, and 10 MLs in (b). (Ref. 29)

growth mode. The In mole fraction x was kept constant (0.15). The dot or island density was about $5 \times 10^{10} cm^{-2}$, roughly an order of magnitude higher than the defect density. The dots with different sizes were obtained by controlling the thickness of $In_xGa_{1-x}N$. Figure 24 presents three sets of PL spectra from the $In_xGa_{1-x}N$ QDs with nominal thickness of 5, 9, 12 MLs (10 MLs for b) at 10 K and room temperature. As the QD size reduces, the PL peak shifts to higher energy with an amount larger than that expected from quantum confinement along. The extra shift is interpreted as the result of QCSE induced by polarization. This interpretation is consistent with the observation that the PL intensity is quenched for large dots, as expected from the QCSE.[65] In their case, the PL was not detected for the dots larger than 10 MLs at room temperature.

Let us estimate the polarization induced electric field in $In_xGa_{1-x}N/GaN(0001)$ QDs. For x=0.15, assuming the dots being fully strained and using the data listed in Table 1, we obtain the interface charge due to the spontaneous polarization as -0.024 C/m^2. The charge due to the piezoelectricity is -0.021 C/m^2, similar to that from the spontaneous polarization. The electric field in QDs induced by these charges is -2.3 MV/cm in QDs if only one interface is considered. The field is doubled if the top and bottom interfaces are both taken into account. This field may be large enough to produce a large shift in the transition energy and quench the oscillator strength for the large dots.

The thermal stability of the PL from $In_xGa_{1-x}N/GaN$ QDs was improved as compared to $In_xGa_{1-x}N$ bulk or QWs,[28,29] but not as good as that observed in GaN QDs in AlN matrix. Typically a few times drop in PL intensity was observed when the temperature was raised from low to room temperature for $In_xGa_{1-x}N$ QDs, [28,29] while a stable PL was reported in GaN/AlN QDs.[14,27] The lower barrier may account for the poorer thermal stability, similar to the cases of $GaN/Al_xGa_{1-x}N$ QDs.[36,37,68] The high dot density may also have a contribution since the thermal tunneling between the adjacent dots may not negligible in this case.

A strong confinement of excitons in $In_xGa_{1-x}N$ QDs was demonstrated by the observation of very sharp and discrete PL lines from a limited number of QDs.[72] The QDs were grown on GaN by MOCVD and their average height/diameter was 4.6/23.4 nm. By patterning a thin aluminum layer with 400 nm square aperture on the sample surface, the detected number of QDs in the PL measurement was reduced to about 20. The measured PL line-width was typically 0.17 meV at 3.5 K (see Figure 25, next page) and limited by the spectral resolution of the measurement system. The separation of these discrete lines is a few meV. When the temperature is raised to 60 K, the width increases to 0.6 meV but is much smaller than the thermal energy kT. The result suggests that the narrow PL lines are really from the strongly localized states. The broadening of the PL spectra commonly observed from a QD assembly is inhomogeneous either from the fluctuation of dot size or alloy composition.

The subband emission similar to the discrete lines described above was also reported from $In_xGa_{1-x}N$ QW laser diodes under room temperature pulse operation.[74] The energy separation is 1-5 meV. Under the room temperature CW operation, a single mode was observed but the mode hopped towards a higher energy with increasing drive current in contrast to that observed in an AlGaInP QW laser. A subband emission was suggested based on the transitions between the ground levels of the $In_xGa_{1-x}N$ QDs of different sizes formed from In-rich regions in the $In_xGa_{1-x}N$ well layers.

The existence of the QD-like structures and its effect on the PL in $In_xGa_{1-x}N$ QWs are the subject of extensive discussion in the past few years.[41,42,74,75,76] The main

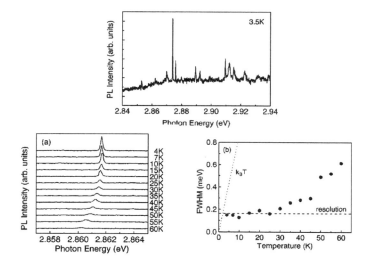

Fig. 25. (a) A high resolution micro-PL spectrum from about 20 numbers of $In_xGa_{1-x}N$ quantum dots measured at 3.5K. (b) The PL spectra of a particular peak taken at different temperatures from 4 to 60 K. (c) The PL linewidth (FWHM) as a function of temperature. Thermal energy is plotted as a dotted line. (Ref. 72)

characteristics of the observed PL spectra are summarized as follows: (1) The luminescence efficiency is high. (2) There is a large Stokes shift from the absorption edge, increasing with In composition. (3) The peak energy increases with temperature at the low temperature region in contrast to the change in the $In_xGa_{1-x}N$ band-gap. (4) The peak energy increases with the excitation density. (5) The PL exhibits different characteristics at high excitation levels. All of the above observations can be well-interpreted by PL from localized states at band-tails, originating from small density of states and long life-time of the trapped carriers. These localized states are believed to be from In-rich or QD-like structures as demonstrated by HRTEM in some samples.[41]

In contrast to the localization effect, Riblet et al.[76] argued that many of the above characteristics may result from the polarization effect in $In_xGa_{1-x}N$ QWs. They investigated the blue shift with excitation intensity for $In_xGa_{1-x}N$ QWs with different well width but the same nominal In mole fraction of 0.25. They found that the maximum shift increases linearly with the well width, from 50 meV for a 1 nm well to 200 meV for a 5 nm well. A similar relation was also found for $GaN/Al_{0.15}Ga_{0.85}N$ QW, but with a shift that is an order of magnitude smaller, see Figure 26 (next page). From these results, they estimated the electric field due to piezoelectricity to be 400-500 kV/cm in $In_{0.25}Ga_{0.75}N$ QWs and 60-70 kV/cm in GaN QW. While this interpretation may be reasonable, the possibility that the PL in $In_xGa_{1-x}N$ QWs is from In-rich dot-like structures is still not ruled out. The effect of the piezoelectricity only amplifies the energy fluctuation for a given dot-size or In mole-fraction distributions.

Surface emitting lasers in $In_xGa_{1-x}N/GaN/Al_xGa_{1-x}N$ structures with $In_xGa_{1-x}N$ QDs were demonstrated recently.[77] The lasing in vertical direction occurs at low temperatures and an ultrahigh gain of 10^5 cm^{-1} in the active region was estimated. The lasing wavelength was 401 nm and the threshold current density was 400 kW/cm^2.

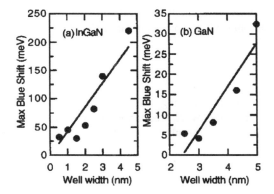

Fig. 26. The maximum blue shift of the PL peak as a function of the well width for (a) $In_{0.25}Ga_{0.75}N$ double QW (with $In_{0.02}Ga_{0.98}N$ barriers) and (b) GaN single QW (with $Al_{0.15}Ga_{0.85}N$ barriers). The maximum blue shift is obtained for an optical excitation intensity of about 500 kW/cm^2. The solid lines are linear adjustments. (Ref. 76)

4. Summary

In the last five years, a great progress has been witnessed on the fabrication and understanding of the III-nitride quantum dots. It is mainly due to the improvement in the growth techniques such as MBE and MOCVD. All three modes, 2D, 3D, and SK were demonstrated in III-nitride epitaxial growth. Fully strained, defect free, vertically correlated GaN QDs in AlN matrix were demonstrated mainly by MBE using SK mode. Strain free GaN QDs in $Al_xGa_{1-x}N$ were also realized using Si anti-surfactant to modify the surface energy and change the growth mode from 2D to 3D. It is demonstrated that the self-assembled GaN QD in AlN or $Al_xGa_{1-x}N$ matrix has a well-defined shape of truncated pyramid with a hexagonal base. The QD assembly with different densities and average sizes can be obtained in a more controllable way. In addition to the GaN QDs, the $In_xGa_{1-x}N$ QDs can also be self-assembled on GaN surface, although the shape, size, and density may be less controllable.

III-nitride QDs show rich and unique optical properties, some of them have been revealed in the last few years mainly by PL investigations. It is realized that the effects of quantum confinement, strain, and polarization are all important to III-nitride QDs and must be properly considered when interpreting any optical spectra. While the quantum confinement and strain in GaN/$Al_xGa_{1-x}N$ QDs commonly induce a blue shift in the transition energy from bulk bandgap, the spontaneous polarization and piezoelectricity may induce a red shift. The latter effects are more significant in the case of fully strained and large size GaN/AlN QDs and may pull the transition energy from bulk GaN bandgap of 3.4 eV down to ~2eV, covering the emission wavelength from blue to orange. A stable PL intensity against thermal perturbation up to room temperature was demonstrated in self-assembled GaN/AlN QDs. The effect of the quantum confinement on the exciton binding energy and electron-phonon coupling were also observed. The transition energy, the thermal stability, and the excitons and the electro-phonon coupling in QDs are all important to applications of III-nitrides in light emitting and detecting devices.

Although great progress has been achieved, the III-nitride QDs grown by any methods today are still far from perfect. As we mentioned in the introduction, fabrication of semiconductor QD assembly with small and uniform size, high density, well-ordered placement, and defect-free materials remain as great challenges, especially in III-Nitride materials.

Acknowledgements

The authors would like to thank T. King for his tireless assistance throughout the laboratory. The VCU portion of this work was funded by grants from AFOSR (Dr. G. L. Witt), NSF (Drs. L. Hess and G. Pomrenke), and ONR (Drs. C. E. C. Wood and Y. S. Park).

References

1. A. D. Yoffe, Adv. Phys. **50**, 1 (2001).
2. D. Bimberg, M. Grundmann, and N. N. Ledentsov, *Quantum Dot Heterostructures* (John Wiley and Sons, Chichester 1999).
3. H. Morkoç, *Nitride Semiconductors and Devices* (Springer Verlag, Heidelberg, 1999).
4. S. N. Mohammad and H. Morkoc, Progress in Quantum Electronics **20**, 361(1996).
5. S. Nakamura and G. Fasol, *The Blue Laser Diode* (Springer-Verlag, Heidelberg, 1997).
6. Y. Arakawa and H. Sakaki, Appl. Phys. Lett. **40**, 939 (1982).
7. S. C. Jain, M. Willander, J. Narayan, R. Van Overstraeten, J. Appl. Phys. **87**, 965 (2000).
8. V. A. Shchukin and D. Bimberg, Rev. Mod. Phys. **71**, 1125 (1999).
9. D. J. Eaglesham and M. Cerullo, Phys. Rev. Lett. **64**, 1943 (1990).
10. P. M. Petroff and S. P. DenBaars, Superlattices and Microstructures **15**, 15 (1994).
11. M. Pinczolits, G. Springholz, and G. Bauer, Appl. Phys. Lett. **73**, 250 (1998).
12. B. Daudin, F. Widmann, G. Feuillet, Y. Samson, M. Arlery, and J. L. Rouviere, Phys. Rev. B **56**, R7069 (1997).
13. B. Daudin, G. Feuillet, F. Widmann, Y. Samson, J. L. Rouviere, N. Pelekanos, and G. Fishman, MRS Symp. Proc. **482**, 205 (1998).
14. F. Widmann, B. Daudin, G. Feuillet, Y. Samson, J. L. Rouviere, and N. Pelekanos, J. Appl. Phys. **83**, 7618 (1998).
15. F. Widmann, J. Simon, B. Daudin, G. Feuillet, J. L. Rouviere, N. Pelekanos, and G. Fishman, Phys. Rev. B **58**, R15989 (1998).
16. J. L. Rouviere, J. Simon, N. Pelekanos, B. Daudin, and G. Feuillet, Appl. Phys. Lett., **75**, 2632 (1999).
17. B. Daudin, F. Widmann, J. Simon, G. Feuillet, J. L. Rouviere, N. Pelekanos, and G. Fishman, MRS Internet J. Nitride Semicond. Res. **4S1**, G9.2 (1999).
18. B. Damilano, N. Grandjean, F. Semond, J. Massies, and M. Leroux, Appl. Phys. Lett., **75**, 962 (1999).
19. X.-Q. Shen, S. Tanaka, S. Iwai, and Y. Aoyagi, Appl. Phys. Lett. **72**, 344 (1998). H. Hirayama, Y. Aoyagi, and S. Tanaka, MRS Internet J. Nitride Semicond. Res. **431**, G9.4 (1999).
20. X. Q. Shen, T. Ide, S. H. Cho, M. Shimizu, S. Hara, H. Okumura, S. Sonoda, and S. Shimizu, Jpn. J. Appl. Phys. **39**, L16 (2000).
21. D. Huang, P. Visconti, K. M. Jones, M. A. Reshchikov, F. Yun, A. A. Baski, T. King, and H. Morkoç, Appl. Phys. Lett. **78**, 4145 (2001).
22. M. Arlery, J. L. Rouviere, F. Widmann, B. Daudin, G. Feuillet, and H. Mariette, Appl. Phys. Lett. **74**, 3287 (1999).
23. G. Feuillet, F. Widmann, B. Daudin, J. Schuler, M. Arlery, J. L. Rouviere, N. Pelekanos, and O. Briot, Mater. Sci. Eng. B **50**, 233 (1997).
24. Q. Xie, Anupam, P. Chen, and N. P. Kobayashi, Phys. Rev. Lett. **75**, 2542 (1995).

25. J. Tersoff, C. Teichert, and M. G. Lagally, Phys. Rev. Lett. **76**, 1675 (1996).
26. J. Gleize, J. Frandon, F. Demangeot, M. Renucci, C. Adelmann, B. Daudin, G. Feuillet, B. Damilano, N. Grandjean, and J. Massies, Appl. Phys. Lett. **77**, 2174 (2000).
27. E. Martinez-Guerrero, C. Adelmann, F. Chabuel, J. Simon, N. T. Pelekanos, G. Mula, B. Daudin, G. Feuillet, and H. Mariette, Appl. Phys. Lett. **77**, 809 (2000).
28. C. Adelmann, J. Simon, G. Feuillet, N. T. Pelekanos, B. Daudin, and G. Fishman, Appl. Phys. Lett. **76**, 1570 (2000).
29. B. Damilano, N. Grandjean, S. Dalmasso, and J. Massies, Appl. Phys. Lett. **75**, 3751 (1999).
30. N. Grandjean and J. Massies, Appl. Phys. Lett., **72**, 1078 (1998).
31. D. Huang, M. A. Reshchikov, F. Yun, T. King, A. A. Baski, and H. Morkoç, Appl. Phys. Lett., in press.
32. P. Visconti, K. M. Jones, M. A. Reshchikov, R. Cingolani, H. Morkoç, and R. J. Molnar, Appl. Phys. Lett. **77**, 3532 (2000).
33. V. Dmitriev, K. Irvine, A. Zubrilov, D. Tsvetkov, V. Nikolaev, M. Jakobson, D. Nelson, and A. Sitnikova, *Gallium Nitride and Related Materials*, edited by R. D. Dupuis, J. A. Edmond, F. A. Ponce, and S. Nakamura (MRS, Pittsburgh, PA, 1996).
34. S. Tanaka, S. Iwai, and Y. Aoyagi, Appl. Phys. Lett. **69**, 4096 (1996).
35. S. Tanaka, H. Hirayama, S. Iwai, and Y. Aoyagi, MRS Symp. Proc. **449**, 135 (1997).
36. P. Ramvall, S. Tanaka, S. Nomura, P. Riblet, and Y. Aoyagi, Appl. Phys. Lett. **75**, 1935 (1999).
37. P. Ramvall, P. Riblet, S. Nomura, Y. Aoyagi, and S. Tanaka, J. Appl. Phys. **87**, 3883 (2000)
38. S. Tanaka, H. Hirayama, Y. Aoyagi, Y. Narukawa, Y. Kawakami, Shizuo Fujita, and Shigeo Fujita, Appl. Phys. Lett. **71**, 1299 (1997).
39. H. Hirayama, S. Tanaka, P. Ramvall, and Y. Aoyagi, Appl. Phys. Lett. **72**, 1736 (1998).
40. K. Tachibana, T. Someya, and Y. Arakawa, Appl. Phys. Lett. **74**, 383 (1999).
41. L. Nistor, H. Bender, A. Vantomme, M. F. Wu, J. Van Landuyt, K. P. O'Donnell, R. Martin, K. Jacobs, and I. Moerman, Appl. Phys. Lett. **77**, 507 (2000).
42. K. P. O'Donnell, R. Martin, P. G. Middleton, Phys. Rev. Lett. **82**, 237 (1999).
43. J. Wang, M. Nozaki, M. Lachab, Y. Ishikawa, R. S. Q. Fareed, T. Wang, M. Hao, and S. Sakai, Appl. Phys. Lett., **75**, 950 (1999).
44. M. Lachab, M. Nozaki, J. Wang, Y. Ishikawa, Q. Fareed, T. Nishikawa, K. Nishino, and S. Sakai, J. Appl. Phys. **87**, 1374 (2000).
45. T. J. Goodwin, V. L. Leppert, H. Risbud, I. M. Kennedy, and H. W. H. Lee, Appl. Phys. Lett., **70**, 3122 (1997).
46. N. Preschilla, S. Major, N. Kumar, I. Samajdar, and R. S. Srinivasa, Appl. Phys. Lett. **77**, 1861 (2000).
47. J. F. Janic and R. L. Well, Chem. Mater. **8**, 2708 (1996).
48. J. L. Coffer, M. A. Johnson, L. Zhang, and R. L. Wells, Chem. Mater. **9**, 2671 (1991).
49. O. I. Micic, S. P. Ahrenkiel, D. Bertram, and A. J. Nozik, Appl Phys. Lett. **75**, 478 (1999).
50. M. Benaissa, K. E. Gonsalves, and S. P. Rangarajan, Appl. Phys. Lett. **71**, 3685 (1997).
51. M. A. Reshchikov, J. Cui, F. Yun, A. A. Baski, M. I. Nathan, and H. Morkoç, Mat. Res. Soc. Symp. Vol. **622**, T4.5 (2000).
52. H. Morkoç, M. A. Reshchikov, K. M. Jones, F. Yun, P. Visconti, M. I. Nathan, and R. J. Molnar, Mat. Res. Soc. Symp. Vol. **639**, G11.2 (2001).
53. D. Das and A. C. Melissinos, *Quantum Mechanics*, (Gordon and Breach Science Publishers, New York, 1986).
54. G. Martin, A. Botchkarev, A. Rockett, and H. Morkoç, Appl. Phys. Lett., **68**, 2541 (1996).
55. V. Y. Davydov, N. S. Averkiev, I. N. Goncharuk, D. K. Delson, I. P. Nikitina, A. S. Polkovnokov, A. N. Smirnov, and M. A. Jacobson, J. Appl. Phys. **82**, 5097(1997).
56. A. Shikanai, T. Azuhata, T. Sota, S. Chichibu, A. Kuramata, K. Horino, and S. Nakamura, J. Appl. Phys. **81**, 417 (1997).
57. H. Amano, K. Hiramatsu, and I. Akasaki, Jpn. J. Appl. Phys. 2 **27**, L1384(1998).
58. W. Shan, R. J. Hauenstein, A. J. Fischer, J. J. Song, W. G. Perry, M. D. Bremser, R. F. Davis, and B. Goldenber, Phys. Rev. B**54**, 13460 (1996).

59. K. Shimada, T. Sota, and K. Suzuki, J. Appl. Phys. **84**, 4951(1998).
60. S. Strite and H. Morkoç, J. Vac. Sci. Technol. B**10**, 1237 (1992).
61. M. S. Shur, A. D. Bykhovski, and R. Gaska, Solid State Electron **44**, 205 (2000).
62. R. Andre, J. Cibert, Le Si Dang, J. Zeman, and M. Zigone, Phys. Rev. B **53**, 6951 (1996).
63. J. F. Nye, *Physical Properties of Crystals*, (Oxford University Press, Oxford, 1985).
64. F. Bernardini, V. Fiorentini, and D. Vanderbilt, Phys. Rev. B**56**, R10024 (1997).
65. E. E. Mendez, G. Bastard, L. L. Chang, L. Esaki, H. Morkoç, and R. Fischer, Phys. Rev. B**26**, 7101 (1982).
66. D. A. B. Miller, D. S. Chemla, T. C. Damen, A. C. Gossard, W. Wiegmann, T. H. Wood, and C. A. Burrus, Phys. Rev. B**32**, 1043 (1985).
67. A. D. Andreev and E. P. O'Reilly, Phys. Rev. B**62**, 15851 (2000).
68. P. Ramvall, S. Tanaka, S. Nomura, P. Riblet, and Y. Aoyagi, Appl. Phys. Lett. **73**, 1104 (1998).
69. A. Petersson, A. Gustafsson, L. Samuelson, S. Tanaka, Y. Aoyagi, Appl. Phys. Lett. **74**, 3513 (1999).
70. W. Shan, B. D. Little, A. J. Fischer, J. J. Song, B. Goldenberg, W. G. Perry, M. D. Bremser, and R. F. Davis, Phys. Rev. B**54**, 16369 (1996).
71. D. C. Reynolds, D. C. Look, W. Kim, Ö. Aktas, A. Botchkarev, A. Salvador, H. Morkoç, and D. N. Talwar, J. Appl. Phys. **80**, 594 (1996).
72. O. Moriwaki, T. Someya, K. Tachibana, S. Ishida, Y. Arakawa, Appl. Phys. Lett. **76**, 2361 (2000).
73. P. Perlin, I. Gorczyca, T. Suski, P. Wisniewski, S. Lepkowski, N. E. Christensen, A. Svane, M. Hansen, S. P. DenBaars, B. Damilano, N. Grandjean, and J. Massies, Phys. Rev. B**64**, 115319 (2001).
74. S. Nakamura, M. Senoh, S. Nagahama, N. Iwasa, T. Yamada, T. Matsushita, Y. Sugimoto, and H. Kiyoku, Appl. Phys. Lett. **70**, 2753 (1997).
75. M. Pophristic, F. H. Long, C. Tran, I. T. Ferguson, and R. F. Karlicek, J. Appl. Phys. **86**, 1114 (1999).
76. P. Riblet, H. Hirayama, A. Kinoshita, A. Hirata, T. Sugano, and Y. Aoyagi, Appl. Phys. Lett. **75**, 2241 (1999).
77. I. L. Krestnikov, A. V. Sakharov, W. V. Lundin, Yu. G. Musikhin, A. P. Kartashova, A. S. Usikov, A. F. Tsatsul'nikov, N. N. Ledentsov, Zh. I. Alferov, P. Soshnikov, E. Hahn, B. Neubauer, A. Rosenbauer, D. Litvinov, D. Gerthsen, A. C. Plaut, A. A. Hoffmann, and D. Bimberg, Semiconductors **34**, 481 (2000).

THEORY OF THRESHOLD CHARACTERISTICS OF QUANTUM DOT LASERS: EFFECT OF QUANTUM DOT PARAMETER DISPERSION

LEVON V. ASRYAN

Ioffe Physico-Technical Institute
26 Polytekhnicheskaya St, St Petersburg 194021, Russia[*]

ROBERT A. SURIS

Ioffe Physico-Technical Institute
26 Polytekhnicheskaya St, St Petersburg 194021, Russia[†]

Gain and threshold current of a quantum dot (QD) laser are analyzed theoretically taking into account the inhomogeneous line broadening caused by fluctuations in QD parameters. Two regimes of QD filling by carriers, nonequilibrium and equilibrium, are identified, depending on temperature, barrier heights and QD size. Critical tolerable parameters of the QD structure, at which lasing becomes impossible to attain, are shown to exist. The minimum threshold current density and optimum parameters are calculated. Violation of charge neutrality in QDs is revealed, which affects significantly the threshold current and its temperature dependence. The gain–current dependence is calculated. The voltage dependences of the electron and hole level occupancies in QDs, gain and current are obtained. The observed temperature dependence of threshold current (constant at low temperatures and rapid increase above a certain temperature) is predicted and explained. Violation of charge neutrality is shown to give rise to the slight temperature dependence of the current component associated with the recombination in QDs. The characteristic temperature T_0 is calculated considering carrier recombination in the optical confinement layer and violation of charge neutrality in QDs. The inclusion of violation of charge neutrality is shown to be critical for the correct calculation of T_0 at low T. T_0 is shown to fall off profoundly with increasing T. Theoretical estimations confirm the possibility of a significant reduction of the threshold currents and enhancement of T_0 of QD lasers as compared with the conventional quantum well (QW) lasers. Longitudinal spatial hole burning is analyzed. Unlike QW lasers, thermally excited escapes of carriers from QDs, rather than diffusion, are shown to control the smoothing-out of the spatially nonuniform population inversion and multimode generation in QD lasers. A decrease in the QD size dispersion is shown not only to decrease the threshold current but to increase considerably the relative multimode generation threshold as well. Concurrent with the increase of threshold current, an increase of the multimode generation threshold is shown to occur with a rise in temperature. Ways to optimize the QD laser, aimed at maximizing the multimode generation threshold, are outlined.

---

[*]Present address: Department of Electrical and Computer Engineering, State University of New York at Stony Brook, Stony Brook, New York 11794-2350, USA
E-mail: asryan@ece.sunysb.edu
[†]E-mail: suris@theory.ioffe.rssi.ru

1. Introduction

Quantum dot (QD) lasers are of particular interest because of the following advantages over the conventional quantum well (QW) lasers: the narrower gain spectra, significantly lower threshold currents and their weaker temperature dependence. A main peculiarity, which the expected advantages of QD lasers spring from, is a very narrow, δ-function-like density of states for the carriers confined in QDs. As a consequence of quantum confinement in all the three dimensions, the energy spectra of carriers are discrete in QDs. Transitions between the electron and hole levels in QDs are analogous to those between the exactly discrete levels of individual atoms. For this reason, QDs have generated much interest as a new class of artificially structured materials with tunable (through varying the compositions and sizes) energies of discrete atomic-like states that are ideal for use in lasers.[1,2] There has been significant progress in fabricating QD lasers [3] and several groups have now demonstrated such devices.[4-19]

The aim of this chapter is to give a theoretical analysis of threshold characteristics of a QD laser taking account of the inhomogeneous line broadening caused by the QD parameter (e.g., size) dispersion. It is not practical to circumvent such a dispersion during the QD structures growth: size fluctuations are inherent in the ensembles of self-organized QDs fabricated by molecular beam epitaxy and metalorganic chemical vapor deposition. This is in contrast to solid state lasers wherein the line broadening is caused by inhomogeneities of the matrix, in which the lasing atoms are inserted, rather than by fluctuations in spectra of atoms. Inhomogeneous broadening is the key factor limiting operating characteristics of a QD laser.[1)]

Though there were theoretical studies of QD lasers,[2)] the question as to how the threshold current and the characteristic temperature depend on the QD size fluctuations, i.e., on the 'degree' of perfection of the structure, remained however unsolved. We analyzed this issue and developed a quantitative picture of the principal processes involved in the laser operation.[23-32] Explicit analytical expressions for the threshold characteristics were derived. Key parameters governing the threshold current and its temperature dependence were revealed.

The chapter is organized as follows. After presenting the basic equations (Section 2), we analyze gain and spontaneous radiative recombination current for equilibrium and nonequilibrium filling of QDs (Section 3). In Section 4, the threshold current density is calculated as a function of structure parameters; optimization of a laser aimed at minimizing the threshold current density is carried out. In Section 5, we discuss violation of the charge neutrality in QDs and its effect on the laser characteristics; gain–current and current–voltage dependences are calculated. The temperature dependence of the threshold current is analyzed in Section 6, where the

---

[1)] Homogeneous broadening presenting in QD lasers as in conventional lasers is not discussed here.
[2)] In Ref.[20], the gain and threshold current have been treated without considering fluctuations in QD sizes. For a normal QD size distribution, the gain was studied in Ref.[21], whereas no consideration has been given to the threshold current. In Ref.[22], a numerical calculation of the threshold current has been reported for a specific value of line broadening.

characteristic temperature of a laser is calculated. In Section 7, longitudinal spatial hole burning is analyzed and the multimode generation threshold is calculated.

2. Basic Equations

The optical confinement layer (OCL) is formed in the field region of the p-n double heterostructure (Fig. 1). The active region, which is a QD array layer (or layers, for details, see Section 4.3.3), is formed in the central part of the OCL along the longitudinal (direction of wave propagation) and lateral directions in the slab-dielectric waveguide. Carrier injection from the cladding layers, which are p- and n-sides of the structure, occurs in the transverse direction (perpendicular to the QD layer).

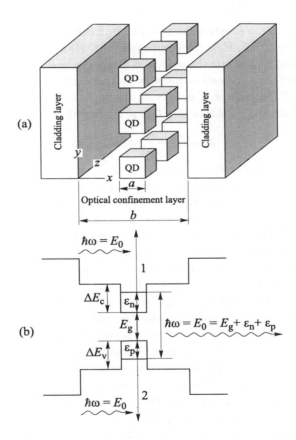

Fig. 1. Schematic (a) and energy band diagram (b) of a QD laser structure. The QDs are not drawn to scale. Arrows 1 and 2 show the transitions of carriers from the quantized energy levels to the continuous-spectrum states in the process of light absorption.

For transitions from a quantized energy level of an electron to that of a hole in a QD, the gain and the spontaneous recombination rate reduced to one QD are

$$g_0(E) = 4\pi^2 \frac{\alpha}{\sqrt{\epsilon}} E |x_{cv}|^2 \rho_{red}(E)(f_n + f_p - 1) \tag{1}$$

$$R_{sp} = \int dE\, 4\alpha \frac{\sqrt{\epsilon} E^3}{\hbar^3 c^2} |x_{cv}|^2 \rho_{red}(E) f_n f_p \tag{2}$$

where $\alpha = e^2/\hbar c$ is the fine structure constant, E is the photon energy, and ϵ is the dielectric constant of the OCL. The gain reduced to one QD is referred to as the material gain.

In a cubic QD,[20] the square of the interband matrix element of the dipole moment formed between an electron and a hole is

$$e^2 |x_{cv}|^2 = e^2 \frac{P^2}{3E^2} = \frac{2}{3} e^2 \left(\frac{\hbar}{2E}\right)^2 \left(\frac{1}{m_c} - \frac{1}{m_0}\right) \frac{E_g(E_g + \Delta_0)}{E_g + 2\Delta_0/3} \tag{3}$$

where m_c is the electron effective mass, m_0 is the free electron mass, Δ_0 is the energy of the spin-orbit splitting, E_g is the band gap of the QD material. The dependence of the matrix element on the QD shape should disappear with increasing QD potential well depth (or QD size). Kane's parameter P enters into (3); we can take advantage of the fact that P depends only weakly on the type of the material.

The reduced density of states in a QD is a δ-function

$$\rho_{red}(E) = \frac{2}{V_0} \delta(E - E_0) \tag{4}$$

where V_0 is the QD volume, $E_0 = E_g + \varepsilon_n + \varepsilon_p$, and ε_n and ε_p are the ground quantized energy levels of an electron and a hole in the conduction and valence bands, respectively, measured from the band edges [Fig. 1(b)]. We restrict our consideration to single electron-level QDs; i.e., the QD sizes and/or the conduction band offset at the QD–OCL heteroboundary (three-dimensional potential well depth for the electron in a QD) are not large enough for the second level to exist. Such limitations on the QD sizes and/or the compositions of QDs and the OCL are reasonable and desired for improvement of the laser characteristics.[33] The point is that the presence of the higher energy levels of holes in rectangular boxlike QDs (provided there are no such electron levels) does not effect significantly the threshold current. The reason is that the radiative transitions from the lowest electron level to the higher hole levels are (at least partly) forbidden in such QDs. For the specific structure with cubic QDs considered in Section 4.3.4, this is the case.

In (1) and (2), $f_{n,p}$ are the occupation probabilities of the levels in a QD. In equilibrium filling of QDs (see below), they are given by the Fermi–Dirac distribution function, $f_{n,p} = \{\exp[(\varepsilon_{n,p} - \mu_{n,p})/T] + 1\}^{-1}$, where $\mu_{n,p}$ are the quasi-Fermi levels for the conduction and valence bands, respectively, measured from the band edges, and the temperature T being measured in terms of energy.

Taking account of the QD size fluctuations, we may write the following equation for the modal gain of a laser:

$$g(E) = Z_L N_S \bar{a}^2 \Gamma \langle g_0(E) \rangle = Z_L N_S \frac{\Gamma}{\bar{a}} \frac{8}{3} \pi^2 \frac{\alpha}{\sqrt{\epsilon}} P^2 \frac{1}{E} \langle (f_n + f_p - 1) \delta(E - E_0) \rangle \tag{5}$$

where Z_L is the number of QD layers, N_S is the surface density of QDs in one layer, $\bar{a}$ is the mean size of QDs, Γ is the mean (per QD layer) optical confinement factor along the transverse direction in the waveguide ($\Gamma \propto \bar{a}$ – see Section 4.3.3), the brackets $\langle ... \rangle$ mean averaging over QDs, $\bar{E}_0 = E_g + \bar{\varepsilon}_n + \bar{\varepsilon}_p$ is the energy of the transition in a mean-sized QD, and $\bar{\varepsilon}_{n,p}$ are the quantized energy levels in a mean-sized QD; $N_S \bar{a}^2$ is the surface fraction of QDs in a QD layer.

Because of the presence of free carriers in the OCL (in the bulk), account must be taken of the radiative recombination there when calculating the total current. We neglect however nonradiative recombination in QDs and in the OCL (we will briefly turn to this question in Section 4.3.4). Thus, the current density is

$$j = Z_L \frac{eN_S}{\tau_{QD}} \langle f_n f_p \rangle + ebBnp \qquad (6)$$

where the first and the second terms are the current densities associated with the spontaneous radiative recombination in QDs and the OCL, respectively. Here b is the thickness of the OCL, n and p are the free electron and hole densities in the OCL, and the reciprocal spontaneous radiative lifetime in QDs is

$$\frac{1}{\tau_{QD}} = \frac{8}{3} \alpha \sqrt{\epsilon} \frac{\bar{E}_0}{\hbar} \left(\frac{P}{\hbar c}\right)^2. \qquad (7)$$

In the strict sense, eq. (7) holds for highly symmetrical (e.g., cubic) QDs with a deep enough confinement potential. In actual low-symmetry QD ensembles, τ_{QD} may depend on the QD shape and the size dependence of τ_{QD} may be stronger than that given by (7). For the specific structure considered in Section 4, $\tau_{QD} = 0.71$ ns.

For nondegenerate materials, the probability of radiative recombination (radiative constant) B (cm^3 s^{-1}) is

$$B = \frac{4}{3}\sqrt{2}\,\pi^{3/2} \alpha \sqrt{\epsilon} \frac{1}{T^{3/2}} \frac{m'^{3/2}_{chh} + m'^{3/2}_{clh}}{m'^{3/2}_c m'^{3/2}_v} E'_g \left(\frac{P'}{c}\right)^2 \qquad (8)$$

where m'_c is the electron effective mass in the OCL; $m'^{3/2}_v = m'^{3/2}_{hh} + m'^{3/2}_{lh}$; m'_{hh} and m'_{lh} are the heavy and light hole effective masses in the OCL, respectively; $m'_{chh} = m'_c m'_{hh}/(m'_c + m'_{hh})$; $m'_{clh} = m'_c m'_{lh}/(m'_c + m'_{lh})$; E'_g and P' are the band gap and Kane's parameter of the material of the OCL, respectively.

In the steady state, the level occupancies in QDs and free carrier densities in the OCL are related by the rate equations $\partial \langle f_{n,p} \rangle / \partial t = 0$:

$$\sigma_n v_n [n(1 - \langle f_n \rangle) - n_1 \langle f_n \rangle] - \frac{1}{\tau_{QD}} \langle f_n f_p \rangle = 0$$

$$\sigma_p v_p [p(1 - \langle f_p \rangle) - p_1 \langle f_p \rangle] - \frac{1}{\tau_{QD}} \langle f_n f_p \rangle = 0 \qquad (9)$$

where $\sigma_{n,p}$ are the cross sections of electron and hole capture into a QD, $v_{n,p}$ are the thermal velocities, and n_1 and p_1 determine intensity of thermal escapes of an electron and a hole from a QD [see (17)]. By analogy with the Sah–Noyce–Shockley–Read trapping centers, the characteristic times of thermally excited escapes are

$$\tau_n^{esc} = \frac{1}{\sigma_n v_n n_1} \qquad \tau_p^{esc} = \frac{1}{\sigma_p v_p p_1}. \qquad (10)$$

The use of the electron and hole level occupancies averaged over the QD ensemble implies that the analysis is carried out in the framework of the mean field approximation. For separate capture of electrons and holes into the multi-level QDs (with equal numbers of electron and hole levels), theory of random population for QDs,[34] which considers carrier capture and recombination as random processes, slightly (by the factor 5/4) enhances the recombination current density in QDs j_{QD} given by the first term in (6). For simultaneous capture of electrons and holes (i.e., of electron–hole pairs), when QDs are considered as neutral, j_{QD} is enhanced by the factor 2.[34] Modifications to j_{QD} given by the theory of random population can be easily accounted for by including the above factors into eq. (7) for the radiative lifetime in QDs. In so doing the equation for j_{QD} will remain unchanged.

In the general case, the relationship between the current density of the QD laser and its gain is given by (5) and (6) in an implicit form. We will analyze these equations for different cases (depending on temperature, QD size fluctuations, and band offsets at the QD–OCL heteroboundary).

3. Gain Spectrum and Spontaneous Radiative Recombination Current

3.1. *Equilibrium filling of QDs (relatively high temperatures and/or shallow potential wells)*

If the characteristic times of thermally excited escapes of carriers from QDs [see (10)] are small compared with the radiative lifetime in QDs

$$\tau_{n,p}^{esc} \ll \tau_{QD} \qquad (11)$$

redistribution of carriers from one QD to another occurs, and quasi-equilibrium distributions are established with the corresponding quasi-Fermi levels determined by the pumping level. As a consequence of such a redistribution, the level occupancies (and numbers of carriers) in various QDs will differ. Averaging over the inhomogeneously broadened ensemble of QDs, we obtain from (5)

$$g(E) = g^{max} (f'_n + f'_p - 1) \frac{w\left[(\bar{E}_0 - E)/(q_n \bar{\varepsilon}_n + q_p \bar{\varepsilon}_p)\right]}{w(0)} \qquad (12)$$

where $f'_{n,p}$ are the values of the Fermi-Dirac distribution functions at the energies $\bar{\varepsilon}_{n,p} - [q_{n,p} \bar{\varepsilon}_{n,p}/(q_n \bar{\varepsilon}_n + q_p \bar{\varepsilon}_p)](\bar{E}_0 - E)$; $f'_n + f'_p - 1$ is the effective population inversion in QDs, which depends on the photon energy E.

The function w is the probability density of relative QD size fluctuations ($a - \bar{a})/\bar{a}$, $\bar{\varepsilon}_{n,p} = \varepsilon_{n,p}(\bar{a})$, and $q_{n,p} = -(\partial \ln \varepsilon_{n,p}/\partial \ln a)|_{a=\bar{a}}$ are numerical constants. If we take expressions for the quantized energy levels in a three-dimensional square well with infinitely high sides as $\varepsilon_{n,p}(a)$, we find $q_{n,p} = 2$.

The effective population inversion tends to its maximum (unity) as $\mu_n + \mu_p \to \infty$. The factor $w\left[(\bar{E}_0 - E)/(q_n\bar{\varepsilon}_n + q_p\bar{\varepsilon}_p)\right]/w(0)$ is at its maximum (unity) at the transition energy in a mean-sized QD $\bar{E}_0$ and decays away from this energy. Hence the quantity $g^{\max}$ in (12) is the maximum possible (saturation) value of the modal gain spectrum peak; it is given as

$$g^{\max} = \frac{\xi}{4}\left(\frac{\bar{\lambda}_0}{\sqrt{\epsilon}}\right)^2 \frac{1}{\tau_{\mathrm{QD}}} \frac{\hbar}{(\Delta\varepsilon)_{\mathrm{inhom}}} \frac{\Gamma}{\bar{a}} Z_{\mathrm{L}} N_{\mathrm{S}} \qquad (13)$$

where ξ is a numerical constant appearing in $w(0)$: $w(0) = \xi/\delta$ ($\xi = 1/\pi$ and $\xi = 1/\sqrt{2\pi}$ for the Lorentzian and Gaussian functions, respectively), δ is the root mean square (RMS) of relative QD size fluctuations, and $\bar{\lambda}_0 = 2\pi\hbar c/\bar{E}_0$.

The inhomogeneous line broadening caused by fluctuations in QD sizes is

$$(\Delta\varepsilon)_{\mathrm{inhom}} = (q_n\bar{\varepsilon}_n + q_p\bar{\varepsilon}_p)\delta. \qquad (14)$$

The maximum (saturation) value of the stimulated emission cross section in a QD averaged over the laser line is

$$\sigma_{\mathrm{QD}}^{\mathrm{m}} = \frac{1}{4}\xi\frac{\hbar}{(\Delta\varepsilon)_{\mathrm{inhom}}}\frac{1}{\tau_{\mathrm{QD}}}\left(\frac{\bar{\lambda}_0}{\sqrt{\epsilon}}\right)^2. \qquad (15)$$

If $(\Delta\varepsilon)_{\mathrm{inhom}}$ and τ_{QD} are taken to represent the Doppler broadening of the line and the intrinsic radiative lifetime for electronic transitions in atoms, respectively, then (15) is exactly converted to the expression for the stimulated emission cross-section in an atom averaged over the gain line in an atomic gas laser.

In view of (11) we can neglect the spontaneous radiative recombination rate in a QD, $\langle f_n f_p\rangle/\tau_{\mathrm{QD}}$, in (9) as compared with the thermally excited escape rates of carriers from a QD, $\sigma_n v_n n_1 \langle f_n\rangle = \langle f_n\rangle/\tau_n^{\mathrm{esc}}$ and $\sigma_p v_p p_1 \langle f_p\rangle = \langle f_p\rangle/\tau_p^{\mathrm{esc}}$, and express the free carrier densities in the OCL in terms of $\langle f_{n,p}\rangle$:

$$n = n_1 \frac{\langle f_n\rangle}{1 - \langle f_n\rangle} \qquad p = p_1 \frac{\langle f_p\rangle}{1 - \langle f_p\rangle}. \qquad (16)$$

Taking into account that $n = N_c \exp[-(\Delta E_{c1} - \mu_n)/T]$, $p = N_v \exp[-(\Delta E_{v1} - \mu_p)/T]$, where $N_{c,v} = 2(m'_{c,v}T/2\pi\hbar^2)^{3/2}$ are the conduction and valence band effective densities of states, and $\Delta E_{c1,v1}$ are the band offsets at the QD–OCL heteroboundary (Fig. 1), we get from (16)

$$n_1 = N_c \exp\left(-\frac{\Delta E_{c1} - \bar{\varepsilon}_n}{T}\right) \qquad p_1 = N_v \exp\left(-\frac{\Delta E_{v1} - \bar{\varepsilon}_p}{T}\right). \qquad (17)$$

Expressions (17) coincide with those for the trapping centers.

With (10) and (17), the condition (11) for the equilibrium filling of QDs becomes

$$T > T_{\mathrm{g}} = \max\left[\frac{\Delta E_{c1} - \bar{\varepsilon}_n}{\ln(\sigma_n v_n N_c \tau_{\mathrm{QD}})}, \frac{\Delta E_{v1} - \bar{\varepsilon}_p}{\ln(\sigma_p v_p N_v \tau_{\mathrm{QD}})}\right]. \qquad (18)$$

The temperature T_g, presenting the boundary between the regions of nonequilibrium and equilibrium fillings of QDs in Fig. 2, is QD size dependent. Firstly, this dependence stems from the such dependence of the electron and hole quantized energies, $\bar{\varepsilon}_n$ and $\bar{\varepsilon}_p$. The processes of carrier capture/escape (and the corresponding cross sections, $\sigma_{n,p}$) and the effects of carrier relaxation are also size dependent in low dimensional systems. Thus, the LA-phonon–mediated relaxation of electrons was shown to become inefficient with decreasing QD size,[35,36] whereas Auger-process–mediated relaxation was shown to be efficient and dominant relaxation mechanism in small-sized QDs.[37] The dependences of the cross sections appear as arguments of logarithmic functions in (18) and, for this reason, affect T_g weaker than such dependences of the quantized energies do.

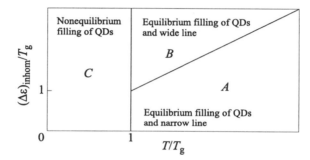

Fig. 2. Different cases of filling of carrier levels in QDs to be realized depending on temperature and QD size fluctuations. T_g is given by (18).

The subsequent analysis should be carried out separately in the cases described in Sections 3.1.1 and 3.1.2.

3.1.1. *Narrow line of the quantized energy distribution (region A in Fig. 2)*

The following inequality holds simultaneously with (18):

$$T > \bar{\varepsilon}_n \delta. \tag{19}$$

We can put w equal to the δ-function. Hence the quantities to be averaged in (6) will be equal to their values at $a = \bar{a}$. With (16), eq. (6) for the current density becomes

$$j = Z_L \frac{eN_S}{\tau_{QD}} \langle f_n \rangle \langle f_p \rangle + ebBn_1 p_1 \frac{\langle f_n \rangle \langle f_p \rangle}{(1 - \langle f_n \rangle)(1 - \langle f_p \rangle)} \tag{20}$$

where $\langle f_{n,p} \rangle = f_{n,p}(\bar{\varepsilon}_{n,p}) = \{[\exp(\bar{\varepsilon}_{n,p} - \mu_{n,p})/T] + 1\}^{-1}$.

In view of (19), the effective population inversion $(f'_n + f'_p - 1)$ in (12) remains practically independent of the transition energy E within the width of the curve for $w[(\bar{E}_0 - E)/(q_n \bar{\varepsilon}_n + q_p \bar{\varepsilon}_p)]$. On putting this factor equal to its value at $E = \bar{E}_0$, we may write $g(E)$ as

$$g(E) = g^{\max} (\langle f_n \rangle + \langle f_p \rangle - 1) \frac{w[(\bar{E}_0 - E)/(q_n \bar{\varepsilon}_n + q_p \bar{\varepsilon}_p)]}{w(0)}. \tag{21}$$

The gain spectrum copies the shape of the curve for $w[(\bar{E}_0 - E)/(q_n \bar{\varepsilon}_n + q_p \bar{\varepsilon}_p)]$ scaled along the vertical axis by a factor equal to the mean population inversion in QDs ($\langle f_n \rangle + \langle f_p \rangle - 1$) [Fig. 3(a)].

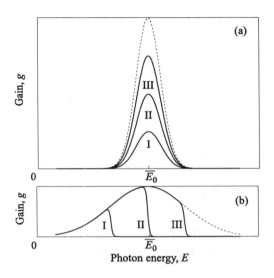

Fig. 3. Gain spectra: (a) for regions A and C in Fig. 2; (b) for region B in Fig. 2. The curves are numbered in ascending order of population inversion in a QD. The dashed curve corresponds to the population inversion equal to unity.

The peak value of the modal gain spectrum, appearing in the threshold condition (28), occurs at $E = \bar{E}_0 = E_g + \bar{\varepsilon}_n + \bar{\varepsilon}_p$:

$$g^m = g^{\max} (\langle f_n \rangle + \langle f_p \rangle - 1). \tag{22}$$

3.1.2. *Wide line of the quantized energy distribution (region B in Fig. 2)*

This is the case in structures with the large dispersion of QD parameters, when

$$T_g < T < \bar{\varepsilon}_n \delta. \tag{23}$$

Analysis of expression (12) for the gain yields: for $E < E_g + \mu_n + \mu_p$, the gain spectrum follows the curve for $w[(\bar{E}_0 - E)/(q_n \bar{\varepsilon}_n + q_p \bar{\varepsilon}_p)]$. In the immediate vicinity of $E = E_g + \mu_n + \mu_p$ the spectral line drops steeply (within the width T) following the drop of the curves for $f'_{n,p}(E)$ [Fig. 3(b)]. At the transition energy

$E = E_g + \mu_n + \mu_p$ (for the corresponding QD, $\varepsilon_n + \varepsilon_p = \mu_n + \mu_p$) the gain is exactly zero.

As $\mu_n + \mu_p$ increases (up to $\bar{\varepsilon}_n + \bar{\varepsilon}_p$), the extreme point of the g dependence on E tends to $\bar{E}_0$ in accordance with the equation $E_m \approx E_g + \mu_n + \mu_p$ (which holds to an accuracy of $T < \bar{\varepsilon}_n \delta$). The maximum gain increases with $\mu_n + \mu_p$ according to $g^m \propto w[(\bar{\varepsilon}_n + \bar{\varepsilon}_p - \mu_n - \mu_p)/(q_n\bar{\varepsilon}_n + q_p\bar{\varepsilon}_p)]$. On further increase in $\mu_n + \mu_p$, E_m and g^m remains unchanged: $E_m = \bar{E}_0$, $g^m \propto w(0) = \xi/\delta$. Thus the gradual 'filling' of the curve for $w[(\bar{E}_0 - E)/(q_n\bar{\varepsilon}_n + q_p\bar{\varepsilon}_p)]$ occurs with increasing pumping level [Fig. 3(b)].

In a limiting case of the severe right-hand inequality in (23) (when $T \ll \bar{\varepsilon}_n\delta$) the current density is

$$j = \frac{eN_S}{\tau_{QD}} \int_{t_m}^{\infty} dt\, w(t) + ebBN_cN_v \exp\left(-\frac{\Delta E_{g1} - \mu_n - \mu_p}{T}\right) \quad (24)$$

where $t_m = \max[(\bar{\varepsilon}_n - \mu_n)/q_n\bar{\varepsilon}_n, (\bar{\varepsilon}_p - \mu_p)/q_p\bar{\varepsilon}_p]$, $\Delta E_{g1} = \Delta E_{c1} + \Delta E_{v1}$ is the band gap difference between the materials of the OCL and QD.

The equations for g^m and j depend on a specific type of the $w[(a-\bar{a})/\bar{a}]$ function. Because of the large QD size dispersion, a fraction of QDs, for which the population inversion condition $E_g + \varepsilon_n + \varepsilon_p \leq E_g + \mu_n + \mu_p$ holds, is small enough. Clearly, the threshold current density in this situation should be considerably higher than in the preceding case.

The situation when inequality $T < \bar{\varepsilon}_p\delta$ holds, instead of the right-hand inequality in (23), is of no particular interest, as the QD size dispersion exceeds the mean size of QDs in that case.

3.2. Nonequilibrium filling of QDs (relatively low temperatures and/or deep potential wells; region C in Fig. 2)

If the radiative lifetime is small compared with the times of thermally excited escapes, which is the case when

$$T < T_g \quad (25)$$

the redistribution of carriers and establishment of quasi-Fermi levels do not occur. Having no time to leave a QD, the carriers recombine there. Since the initial numbers of carriers injected into various QDs are the same, the level occupancies are also the same there. The contribution of each QD to the lasing is the same. A mathematical treatment akin to that used for the deduction of (12) shows that the equations for the gain spectrum and maximum gain are identical to those in the case corresponding to region A in Fig. 2 [see (21) and (22)]. The only difference between them is the following: the level occupancies in a mean-sized QD, $\langle f_{n,p} \rangle = f_{n,p}[\varepsilon_{n,p}(\bar{a})]$, and the level occupancies common to all QDs, $f_{n,p}$, enter into these equations in the former case and in the case under consideration, respectively.

Neglecting the thermally excited escape rates as compared with the spontaneous recombination rate in (9), we can express the free carrier densities in the OCL as

$$n = \frac{1}{\sigma_n v_n \tau_{QD}} \frac{f_n f_p}{1 - f_n} \qquad p = \frac{1}{\sigma_p v_p \tau_{QD}} \frac{f_n f_p}{1 - f_p}. \qquad (26)$$

With (26), eq. (6) for the current density becomes

$$j = Z_L \frac{e N_S}{\tau_{QD}} f_n f_p + \frac{ebB}{\sigma_n \sigma_p v_n v_p \tau_{QD}^2} \frac{f_n^2 f_p^2}{(1 - f_n)(1 - f_p)}. \qquad (27)$$

4. Threshold Current Density and Optimization of a Laser

At the lasing threshold, the modal gain spectrum peak is equal to the total losses

$$g^m = \beta + \beta^{ex} \qquad (28)$$

where β is the loss coefficient in the waveguide, and β^{ex} is the effective absorption coefficient for the process of carrier photoexcitation from the quantized energy levels in QDs to the continuous-spectrum states. The injection current density, at which (28) holds, is the threshold current density of a laser j_{th}.

With the g^m dependence on the level occupancies in QDs, we can determine the values of these latter that satisfy (28) and then, substituting into the expression for the current density, obtain j_{th}.

Since the initial expressions for the maximum gain and current density do not depend on a specific type of the $w[(a - \bar{a})/\bar{a}]$ function in the cases corresponding to regions A and C in Fig. 2, the expressions for j_{th} will also show no such dependence. A detailed analysis of j_{th} will be performed for these cases. [In the case corresponding to region B in Fig. 2, to obtain j_{th}, we must eliminate the quasi-Fermi levels $\mu_{n,p}$ from the expressions for the gain and current density by using the threshold condition (28). Since these expressions depend on a specific type of the $w[(a - \bar{a})/\bar{a}]$ function, there is no way of obtaining in an explicit form the expression for j_{th} for an arbitrary shape of the $w[(a - \bar{a})/\bar{a}]$ curve. However, this case is of little importance in view of the fact that j_{th} is high as compared to that in the other cases discussed.]

4.1. Carrier photoexcitation from QDs to the continuum

QD-carrier excitation by photons emitted during the laser operation [Fig. 1(b)] is analogous to the process of free-carrier absorption in a bulk material. However, in contrast to the latter process, no third particle must participate [together with an electron (hole) and a photon] in the absorption event in a QD to satisfy the momentum conservation law. The boundaries of a three-dimensional QD well act as the "third particle": the required momentum is transferred to the carrier owing to its interaction with them. To estimate the absorption coefficient for the photoexcitation process, we used the simplest approach which takes no account of the complex structure of the valence band and of the interaction between c- and v-bands. In this rough approximation, we have $\beta^{\text{ex}} = \beta_n^{\text{ex}} + \beta_p^{\text{ex}}$, where [25,32]

$$\beta_{n,p}^{\text{ex}} = 64\sqrt{2}\,\pi^3 Z_L N_S \bar{a}^2 \frac{\Gamma}{\bar{a}} \frac{\alpha}{\sqrt{\epsilon}} \frac{\Delta E_{c1,v1}}{\bar{E}_0} \left(\frac{\hbar^2}{m_{c,v}\bar{a}^2 \bar{E}_0}\right)^{5/2} \langle f_{n,p}\rangle. \tag{29}$$

The effective absorption coefficient is much less than β. For the representative values of N_S and Γ of the order of 10^{11} cm^{-2} and several percent, respectively (see below), it is less than or of the order of 0.1 cm^{-1}, whereas the typical values of losses β are of the order of 10 cm^{-1}. Thus, it is essential to take account of light absorption in the process of carrier photoexcitation from the QD levels to the continuous-spectrum states only at very low total losses ($\beta < 1$ cm^{-1}), i.e. for long-cavity lasers. For this reason, we neglect β^{ex} in (28).

4.2. Critical tolerable parameters

As shown above, the modal gain saturates at $g^{\max}$. This value is attained when the electron and hole level occupancies are at their maximum, that is, both unity together. With (22) and $\beta = (1/L)\ln(1/R)$, where L is the cavity length and R is the facet reflectivity, the threshold condition (28) may be written as

$$g^{\max}(\langle f_n\rangle + \langle f_p\rangle - 1) = \frac{1}{L}\ln\frac{1}{R}. \tag{30}$$

Thus we obtain immediately a necessary condition for the lasing to be possible to attain: $g^{\max} \geq (1/L)\ln(1/R)$. With eq. (13) for $g^{\max}$, we get the following inequality for tolerable values of structure parameters:

$$\frac{4}{\xi}\left(\frac{\sqrt{\epsilon}}{\lambda_0}\right)^2 \tau_{\text{QD}} \frac{a}{\Gamma} \frac{(q_n\varepsilon_n + q_p\varepsilon_p)\delta}{\hbar} \frac{1}{Z_L} \frac{1}{N_S} \frac{1}{L}\ln\frac{1}{R} \leq 1. \tag{31}$$

Inequality (31) relates tolerable values of structure parameters to each other. For the specific GaInAsP/InP structure lasing at $1.55\,\mu$m (Section 4.3.4), the three-dimensional region of tolerable N_S, δ, and L is the region above the surface in Fig. 4.

The critical tolerable value of each of the structure parameters is related to the given values of the other parameters. The relation is given by the limiting case of equality in (31) (the surface in Fig. 4). These critical values are the minimum surface density of QDs $N_S^{\min}$, maximum RMS of relative QD size fluctuations $\delta^{\max}$, and minimum cavity length $L^{\min}$ (see also Section 4.3.3):

$$N_S^{\min} = \frac{4}{\xi}\left(\frac{\sqrt{\epsilon}}{\lambda_0}\right)^2 \tau_{\text{QD}} \frac{(q_n\varepsilon_n + q_p\varepsilon_p)\delta}{\hbar} \beta \frac{\bar{a}}{\Gamma} \frac{1}{Z_L} \tag{32}$$

$$\delta^{\max} = \frac{\xi}{4}\left(\frac{\lambda_0}{\sqrt{\epsilon}}\right)^2 \frac{1}{\tau_{\text{QD}}} \frac{\Gamma}{a} \frac{\hbar}{(q_n\varepsilon_n + q_p\varepsilon_p)} \left(\ln\frac{1}{R}\right)^{-1} L Z_L N_S \tag{33}$$

$$L^{\min} = \frac{4}{\xi}\left(\frac{\sqrt{\epsilon}}{\lambda_0}\right)^2 \tau_{\text{QD}} \frac{a}{\Gamma} \frac{(q_n\varepsilon_n + q_p\varepsilon_p)\delta}{\hbar} \frac{1}{Z_L} \frac{1}{N_S}\ln\frac{1}{R}. \tag{34}$$

The more perfect the structure (the less δ) or the longer the cavity, the less is $N_S^{\min}$. The denser the QD ensemble (the greater N_S) or the longer the cavity, the greater is $\delta^{\max}$. The more perfect the structure or the denser the QD ensemble, the less is $L^{\min}$.

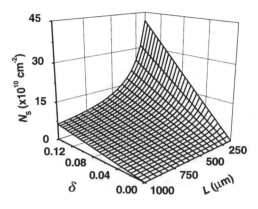

Fig. 4. 3-D region (above the surface) of tolerable values of the surface density of QDs, RMS of relative QD size fluctuations, and cavity length [see (31)].

With the critical tolerable parameters, the threshold condition becomes

$$\langle f_n \rangle + \langle f_p \rangle - 1 = \frac{N_S^{\min}}{N_S}. \tag{35}$$

Equation (35) determines the population inversion in the mean-sized QD required for the lasing to occur at a given surface density of QDs N_S. The ratios $\delta/\delta^{\max}$ or $L^{\min}/L$ can be equivalently inserted into the right-hand side of (35) instead of $N_S^{\min}/N_S$.

As one of structure parameters is close to its critical tolerable value, both $\langle f_n \rangle$ and $\langle f_p \rangle$ tend to unity together — the electron and hole levels are fully occupied. This demands infinitely high free-carrier densities in the OCL [see (16) and (26)]. As a result, the threshold current density j_{th} increases infinitely [see (20) and (27) and Figs. 5 and 6].

4.3. *Equilibrium filling of QDs and narrow line of the quantized energy distribution*

With (20) and (35), we obtain the threshold current density as

$$j_{\text{th}} = Z_L \frac{eN_S^{\min}}{\tau_{\text{QD}}} + ebBn_1p_1$$

$$+ Z_L \frac{eN_S^{\min}}{\tau_{\text{QD}}} \frac{(1 - \langle f_n \rangle)(1 - \langle f_p \rangle)}{\langle f_n \rangle + \langle f_p \rangle - 1} + ebBn_1p_1 \frac{\langle f_n \rangle + \langle f_p \rangle - 1}{(1 - \langle f_n \rangle)(1 - \langle f_p \rangle)}. \tag{36}$$

The level occupancies in QDs are related by (35). The second equation relating $\langle f_n \rangle$ to $\langle f_p \rangle$ should be derived from the solution of the problem for the electrostatic field distribution across the junction (Section 5).

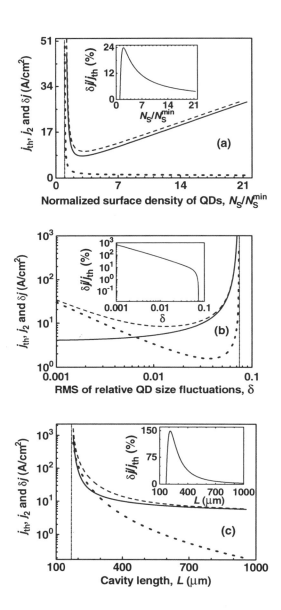

Fig. 5. Threshold current densities for the main (the solid curves) and the next (the dashed curves) longitudinal modes and the multimode generation threshold (the dotted curves) against normalized surface density of QDs (a), RMS of relative QD size fluctuations (b), and cavity length (c). The insets show the relative multimode generation threshold (see Section 7).

4.3.1. Dependence on the surface density of QDs

In the special case of a symmetric structure ($\mu_n - \bar{\varepsilon}_n = \mu_p - \bar{\varepsilon}_p$), this relation reads as $\langle f_n \rangle = \langle f_p \rangle$. Together with (35), this yields

$$\langle f_{n,p} \rangle = \frac{1}{2}\left(1 + \frac{N_S^{\min}}{N_S}\right). \tag{37}$$

The threshold current density is

$$j_{\text{th}}(N_S) = Z_L \frac{1}{4} \frac{eN_S^{\min}}{\tau_{QD}} \left(1 + \frac{N_S^{\min}}{N_S}\right)^2 \frac{N_S}{N_S^{\min}} + ebBn_1p_1 \frac{(1+N_S^{\min}/N_S)^2}{(1-N_S^{\min}/N_S)^2} \tag{38}$$

where $N_S > N_S^{\min}$. The dependence of j_{th} on N_S is nonmonotonic [Fig. 5(a)]. Whereas the recombination current in the OCL [the second term in (20) and (38)] increases infinitely with $N_S \to N_S^{\min}$, the recombination current in QDs [the first term in (38)] behaves similarly with $N_S \to \infty$. This also has a simple explanation: as $N_S \to \infty$, $\langle f_{n,p} \rangle$ tend not to zero but to a finite value, which is 1/2 [see (37)]. Thus, to ensure the lasing, there must be, on the average, one electron and one hole for each QD. To do this would require an infinitely high pumping when $N_S \to \infty$.

With $N_S \to \infty$, the curve for j_{th} approaches asymptotically the straight line [Figs. 5(a) and 15] depicting the transparency current density, i.e., the current density at the inversion threshold $\langle f_n \rangle + \langle f_p \rangle - 1 = 0$:

$$j_{\text{tr}} = Z_L \frac{1}{4} \frac{eN_S}{\tau_{QD}} + ebBn_1p_1. \tag{39}$$

To attain lasing at high N_S, only a very small positive population inversion in QDs is necessary to create, hence the proximity of j_{th} to j_{tr} at such values of N_S.

4.3.2. Dependence on the QD size dispersion and the cavity length

Equation (38) will also present j_{th} as an explicit function of δ (or L) if the ratio $N_S^{\min}/N_S$ is replaced by $\delta/\delta^{\max}$ (or $L^{\min}/L$). Thus we obtain

$$j_{\text{th}}(\delta) = \frac{1}{4} \frac{eN_S}{\tau_{QD}} \left(1 + \frac{\delta}{\delta^{\max}}\right)^2 + ebBn_1p_1 \frac{\left(1 + \frac{\delta}{\delta^{\max}}\right)^2}{\left(1 - \frac{\delta}{\delta^{\max}}\right)^2} \quad (\delta < \delta^{\max}). \tag{40}$$

As seen from (40), j_{th} decreases and approaches the transparency current density with $\delta \to 0$ or $L \to \infty$ [Fig. 5(b,c)]. As already mentioned, $j_{\text{th}} \to \infty$ as $\delta \to \delta^{\max}$ or $L \to L^{\min}$ [Fig. 5(b,c)]. Such an infinite increase in j_{th} with the QD size dispersion approaching a certain value was observed experimentally.[18]

4.3.3. Optical confinement factor and dependence on the OCL thickness

From (38) and (32) it follows reasonably that the larger the optical confinement factor per QD layer Γ, the lower is j_{th}. Because of descending (from the center of the OCL) behavior of the light intensity distribution across the transverse direction, Γ is maximum in the case of one QD layer ($Z_\mathrm{L} = 1$) located in the center of the OCL. Thus, one is inclined to think that the best (aiming at minimizing j_th) plan of QD siting is to arrange them within one layer in the center of the OCL (Fig. 1). This conclusion stems from the assumption that to attain the modal gain sufficiently high to overcome the losses, the density of the QD arrangement in the layer can be made as high as desired [see (43) below]. In actuality, however, such is not the case. The matter is that for closely spaced QDs, the electron tunneling between QDs must not be neglected. The tunneling effect, resulting in the energy level splitting in individual QDs and hence in the laser line broadening, will not enable QDs to be considered as separate. Thus, at large losses (and/or at large QD size fluctuations), elimination of this cause of line broadening will require QD siting in more than one layer (the requisite number of layers will be controlled by the specific values of β and δ). For the β and δ values considered here ($\beta \leq 40\,\mathrm{cm}^{-1}$, $\delta \leq 0.2$), the QD arrangement in one layer would suffice to attain the excess of the gain over the losses. In so doing the separations between neighboring QDs will well exceed those at which the tunneling will be significant.

On using the expression for the electric field in a symmetric three-layer slab waveguide, the following expression for the optical confinement factor in the QD layer of thickness $\bar{a}$ is obtained:

$$\Gamma = \frac{\bar{a}}{b/2 + 1/\gamma} \qquad (41)$$

where b is the thickness of the OCL, $\gamma = \sqrt{K^2 - \epsilon' \bar{E}_0^2/\hbar^2 c^2}$, ϵ' is the dielectric constant of the cladding layers, K is the propagation constant to be found from the eigenvalue equation $\tan(\chi b/2) = \gamma/\chi$, $\chi = \sqrt{\epsilon \bar{E}_0^2/\hbar^2 c^2 - K^2}$. The quantities K, χ and γ depend on b (γ increases from $(1/2)(\epsilon - \epsilon')(\bar{E}_0^2/\hbar^2 c^2)b$ at small b to $\sqrt{\epsilon - \epsilon'}\bar{E}_0/\hbar c$ at large b).

The dependence of Γ on b is plotted in universal form in Fig. 6(a). The abscissa and ordinate are the normalized thickness of the OCL and normalized optical confinement factor,

$$\frac{b}{\bar{\lambda}_0/(2\pi\sqrt{\epsilon - \epsilon'})} \quad \text{and} \quad \frac{\Gamma}{2\pi\sqrt{\epsilon - \epsilon'}\bar{a}/\bar{\lambda}_0}.$$

The maximum value of the normalized optical confinement factor is $z^3/\sqrt{1+z^2} \approx 0.406$ ($z \approx 0.805$ is the root of the equation $1 + z\arctan z = 1/z^2$). The maximum occurs at a normalized thickness to $2\sqrt{1+z^2}\arctan z \approx 1.739$.

Inasmuch as the dependence of Γ on b is nonmonotonic, the dependence of j_th on b is also nonmonotonic [Fig. 6(b)]. In Fig. 6(b), b_min and b_max are the minimum and maximum tolerable thicknesses of the OCL, respectively (see Section 4.2 for other critical parameters). They determine the lower and upper bounds of the OCL thickness interval wherein the threshold condition (28) holds. These parameters are the roots of the equation

$$\Gamma(b) = \frac{4}{\xi} \left(\frac{\sqrt{\epsilon}}{\bar{\lambda}_0} \right)^2 \tau_{QD} \frac{(\Delta\varepsilon)_{\text{inhom}}}{\hbar} \beta \frac{\bar{a}}{N_S} \quad (42)$$

derived from the condition $\langle f_{n,p} \rangle = 1$ or $N_S^{\min} = N_S$ [see (32) and (37)]. For $b_{\min} < b < b_{\max}$, $\langle f_{n,p} \rangle < 1$. For $N_S < N_{S,0}^{\min}$, where

$$N_{S,0}^{\min} = \frac{1}{\pi} \frac{2}{\xi} \frac{\sqrt{1+z^2}}{z^3} \frac{\epsilon}{\sqrt{\epsilon - \epsilon'}} \tau_{QD} \frac{\beta}{\bar{\lambda}_0} \frac{(\Delta\varepsilon)_{\text{inhom}}}{\hbar} \approx \frac{1.569}{\xi} \frac{\epsilon}{\sqrt{\epsilon - \epsilon'}} \tau_{QD} \frac{\beta}{\bar{\lambda}_0} \frac{(\Delta\varepsilon)_{\text{inhom}}}{\hbar} \quad (43)$$

eq. (42) possesses no solution. Thus, $N_{S,0}^{\min}$ presents the minimum value of the dependence of $N_S^{\min}$ on the OCL thickness b.

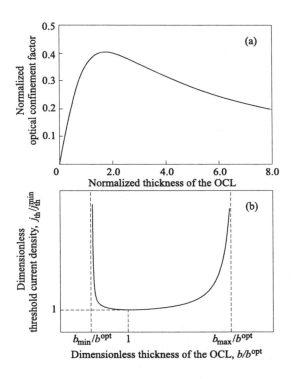

Fig. 6. (a) Universal dependence of the normalized optical confinement factor on the normalized thickness of the OCL. (b) Dimensionless threshold current density $j_{\text{th}}/j_{\text{th}}^{\min}$ against dimensionless thickness of the OCL b/b^{opt}. b^{opt} is evaluated from (46). $b_{\min}$ and $b_{\max}$ are the roots of (42).

With the OCL thickness approaching its minimum or maximum critical tolerable values $b_{\min}$ or $b_{\max}$, the level occupancies $\langle f_{n,p} \rangle$ tend to unity, this resulting in an infinite increase of the recombination current density in the OCL [Fig. 6(b)].

The OCL thickness b^{opt}, at which the minimum of j_{th} occurs, is distinct from that at which the optical confinement factor is a maximum. This fact owes to the direct dependence of the recombination current density in the OCL on b [in addition to the nonmonotonic implicit dependence of j_{th} on b via a similar dependence of Γ entering in eq. (32) for $N_S^{\min}$].

In the general case of $\langle f_n \rangle \neq \langle f_p \rangle$, the character of the dependence of j_{th} on structure parameters is identical to that for a symmetric structure.

4.3.4. Optimization of a laser

We now return to eq. (36), which gives j_{th} for an arbitrary relationship between $\langle f_n \rangle$ and $\langle f_p \rangle$, and are coming to the minimization of j_{th} with respect to N_S and b. It is an easy matter to see that whatever the specific type of the function $\varphi(N_S) = (1 - \langle f_n \rangle)(1 - \langle f_p \rangle)/(\langle f_n \rangle + \langle f_p \rangle - 1)$ is, the term $[(eN_S^{\min}/\tau_{\text{QD}})\varphi + (ebBn_1p_1)/\varphi]$ appearing in (36) is a minimum [being equal to $2\sqrt{(eN_S^{\min}/\tau_{\text{QD}})(ebBn_1p_1)}$] when $\varphi = \sqrt{(ebBn_1p_1)/(eN_S^{\min}/\tau_{\text{QD}})}$. This yields immediately the following equation for $j_{\text{th}}^{\min} = j_{\text{th}}(N_S^{\text{opt}}, b^{\text{opt}})$:

$$j_{\text{th}}^{\min} = \left\{ \left[\frac{eN_S^{\min}(b^{\text{opt}})}{\tau_{\text{QD}}} \right]^{1/2} + (eb^{\text{opt}} Bn_1p_1)^{1/2} \right\}^2$$

$$= eb^{\text{opt}} Bn_1p_1 \left\{ 1 + \left[\frac{\bar{a}}{b^{\text{opt}} \Gamma(b^{\text{opt}})} s \right]^{1/2} \right\}^2 \quad (44)$$

where the dimensionless parameter

$$s = \frac{(4/\xi)(\sqrt{\epsilon}/\bar{\lambda}_0)^2 \, ((\Delta\varepsilon)_{\text{inhom}}/\hbar) \, \beta}{Bn_1 p_1} \quad (45)$$

represents the ratio of the stimulated transition rate in QDs at the lasing threshold to the spontaneous transition rate in the OCL at the transparency threshold.

In (44), b^{opt} is the optimum thickness of the OCL and it should be obtained by minimization of $[(eN_S^{\min}/\tau_{\text{QD}})^{1/2} + (ebBn_1p_1)^{1/2}]^2$, where $N_S^{\min}$ depends on b [see (32)]. For a QD layer located in the center of the OCL [when Γ is given by (41)], b^{opt} is to be found from

$$\left(\frac{\gamma^{\text{opt}} b^{\text{opt}}}{1 + \gamma^{\text{opt}} b^{\text{opt}}/2} \right)^{1/2} \left[\frac{(\chi^{\text{opt}}/\gamma^{\text{opt}})^2}{1 + \gamma^{\text{opt}} b^{\text{opt}}/2} - 1 \right] = \frac{2}{\sqrt{s}} \quad (46)$$

where $\gamma^{\text{opt}} = \gamma(b^{\text{opt}})$, $\chi^{\text{opt}} = \chi(b^{\text{opt}})$.

Thus, for any device design, i.e., for an arbitrary relationship between $\langle f_n \rangle$ and $\langle f_p \rangle$ and for an arbitrary location of the QD layer in the OCL, the minimum threshold current density is given in universal form by (44). The minimum is reached when the condition

$$\frac{(1 - \langle f_n \rangle^{\text{opt}})(1 - \langle f_p \rangle^{\text{opt}})}{\langle f_n \rangle^{\text{opt}} + \langle f_p \rangle^{\text{opt}} - 1} = \left[\frac{b^{\text{opt}} \Gamma(b^{\text{opt}})}{\bar{a}} \frac{1}{s} \right]^{1/2} \quad (47)$$

is satisfied [see (36)]. Equation (47) relates the optimum $\langle f_n \rangle$ to optimum $\langle f_p \rangle$.

Given (35) and the design-dependent relationship between $\langle f_n \rangle$ and $\langle f_p \rangle$, the optimum surface density of QDs, N_S^{opt}, can be evaluated from (47). Hence it follows that, in contrast to j_{th}^{min} and b^{opt}, N_S^{opt} is design-dependent.

For a symmetric structure [see (37)],

$$\langle f_{n,p}\rangle^{opt} = 1 - \left\{1 + \left[1 + \left(\frac{\bar{a}}{b^{opt}\Gamma(b^{opt})}s\right)^{1/2}\right]^{1/2}\right\}^{-1} \tag{48}$$

$$\frac{N_S^{opt}}{N_S^{min}(b^{opt})} = 1 + 2\left(\frac{b^{opt}\Gamma(b^{opt})}{\bar{a}}\frac{1}{s}\right)^{1/2} + \left\{\left[1 + 2\left(\frac{b^{opt}\Gamma(b^{opt})}{\bar{a}}\frac{1}{s}\right)^{1/2}\right]^2 - 1\right\}^{1/2}. \tag{49}$$

Such a structure with a QD layer in the center of the OCL is considered in this section.

The following dimensionless normalized quantities are plotted in universal form in Fig. 7 versus the dimensionless parameter s: optimum thickness of the OCL $(2\pi/\bar{\lambda}_0)\sqrt{\epsilon - \epsilon'}b^{opt}$ (a), optical confinement factor $\Gamma(b^{opt})/[(2\pi/\bar{\lambda}_0)\sqrt{\epsilon - \epsilon'}\bar{a}]$ (b), optimum surface density of QDs (c)

$$\frac{N_S^{opt}/\tau_{QD}}{[\bar{\lambda}_0/(2\pi\sqrt{\epsilon - \epsilon'})]Bn_1p_1}$$

and minimum threshold current density (d)

$$\frac{j_{th}^{min}/e}{[\bar{\lambda}_0/(2\pi\sqrt{\epsilon - \epsilon'})]Bn_1p_1}.$$

The quantity $[\bar{\lambda}_0/(2\pi\sqrt{\epsilon - \epsilon'})]Bn_1p_1$, to which N_S^{opt}, $N_S^{min}(b^{opt})$, and j_{th}^{min} are normalized, is the spontaneous radiative recombination flux in the OCL at the transparency threshold and at its thickness equal to $\bar{\lambda}_0/(2\pi\sqrt{\epsilon - \epsilon'})$. The quantities N_S^{opt}/τ_{QD} and $N_S^{min}(b^{opt})/\tau_{QD}$ are the radiative recombination fluxes in QDs at the surface density of QDs equal to N_S^{opt} and $N_S^{min}(b^{opt})$, respectively, and at the maximum level occupancy in QDs.

For large QD size fluctuations and/or large losses [$s \gg 1$ – see (45)], $\Gamma(b^{opt})$ and b^{opt} tend to the maximum of the optical confinement factor and to the thickness of the OCL at which it occurs (see Figs. 6 and 7). The reason is that when the second term in (38) is negligibly small as compared to the first one, the minimization of j_{th} with respect to b reduces to finding the maximum of Γ. The optimum surface density of QDs is close to the minimum one (at the OCL thickness $b = b^{opt}$); the asymptotic expression for N_S^{opt} is coincident with expression (43) for $N_{S,0}^{min}$. The level occupancies in QDs $\langle f_{n,p}\rangle^{opt}$ are close to unity. The minimum threshold current density is

$$j_{th}^{min} \approx 1.569 \frac{1}{\xi} e \frac{\epsilon}{\sqrt{\epsilon - \epsilon'}} \frac{\beta}{\bar{\lambda}_0} \frac{(\Delta\varepsilon)_{inhom}}{\hbar}. \tag{50}$$

Thus, for relatively large QD size fluctuations, $N_{\rm S}^{\rm opt}$ and $j_{\rm th}^{\rm min}$ are linear with the inhomogeneous line broadening $(\Delta\varepsilon)_{\rm inhom}$ and the total losses β.

For small QD size fluctuations and/or small losses ($s \ll 1$), the optimum parameters $b^{\rm opt}$, $\Gamma(b^{\rm opt})$, $N_{\rm S}^{\rm opt}$, and the minimum threshold current density $j_{\rm th}^{\rm min}$ are proportional to $\sqrt{(\Delta\varepsilon)_{\rm inhom}}$ and $\sqrt{\beta}$ (Figs. 7 and 8):

$$N_{\rm S}^{\rm opt} = \frac{4+3\sqrt{2}}{\pi}\tau_{\rm QD}\left[\frac{1}{\xi}\frac{\epsilon}{\epsilon-\epsilon'}Bn_1p_1\beta\frac{(\Delta\varepsilon)_{\rm inhom}}{\hbar}\right]^{1/2} \qquad (51)$$

$$j_{\rm th}^{\rm min} = 4\sqrt{2}\frac{1}{\pi}e\left[\frac{1}{\xi}\frac{\epsilon}{\epsilon-\epsilon'}Bn_1p_1\beta\frac{(\Delta\varepsilon)_{\rm inhom}}{\hbar}\right]^{1/2}. \qquad (52)$$

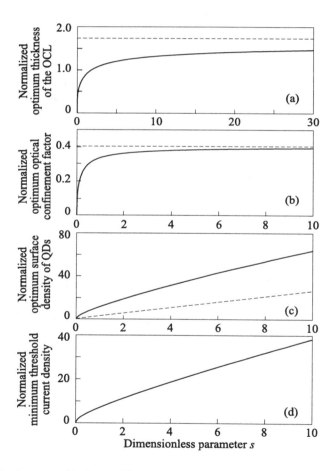

Fig. 7. Dimensionless normalized quantities as universal functions of the dimensionless parameter s. (a) Optimum thickness of the OCL. The horizontal dashed line depicts the asymptote for $s \to \infty$. (b) Optimum optical confinement factor. The horizontal dashed line depicts the asymptote for $s \to \infty$. (c) Optimum surface density of QDs. The dashed curve depicts the minimum surface density of QDs (at $b = b^{\rm opt}$). (d) Minimum threshold current density.

It will be seen from (51) and (52) that at relatively high temperatures ($s \ll 1$), the minimum threshold current density and optimum surface density of QDs strongly depend on the temperature and conduction and valence band offsets at the QD–OCL heteroboundary (i.e., on the potential well depths for carriers in a QD)

$$j_{th}^{min}, N_S^{opt} \propto \exp\left(-\frac{\Delta E_{g1} - \bar{\varepsilon}_n - \bar{\varepsilon}_p}{2T}\right). \qquad (53)$$

As the temperature decreases and/or as the quantity $\Delta E_{g1} - \bar{\varepsilon}_n - \bar{\varepsilon}_p = (\Delta E_{c1} - \bar{\varepsilon}_n) + (\Delta E_{v1} - \bar{\varepsilon}_p)$ increases, the free carrier densities in the OCL and, consequently, the stray current due to the recombination in the OCL decrease. This results in decreased j_{th}^{min} and N_S^{opt} and in weakening of their temperature dependences. It is reasonable that for low T ($s \gg 1$), when the spontaneous radiative recombination in the OCL is not a factor, the dependence of j_{th}^{min} and N_S^{opt} on T and $\Delta E_{c,v}$ disappears [see (43) and (50)]. For $T < T_g$ [see eq. (18) for T_g], when the QD filling is nonequilibrium (region C in Fig. 2), the threshold current density is temperature-independent (see Section 4.4). We will turn to the T-dependences of the laser characteristics in Section 6.

So, (44)–(49) determine the minimum threshold current density and optimum parameters of the structure as universal functions of the dimensionless parameter s controlled by the inhomogeneous line broadening $(\Delta \varepsilon)_{inhom}$, total losses β, temperature, and band offsets at the QD–OCL heteroboundary $\Delta E_{c1,v1}$. By reference to the universal dependences of the dimensionless normalized quantities on s shown in Fig. 7, it is an easy matter to obtain the corresponding dimensional quantities as functions of each of the factors controlling s by itself at the fixed remaining factors.

As an illustration we considered a $Ga_xIn_{1-x}As_yP_{1-y}/InP$ heterostructure lasing near 1.55 μm. The cladding layers, OCL, and QDs are InP, $Ga_{0.21}In_{0.79}As_{0.46}P_{0.54}$, and $Ga_{0.47}In_{0.53}As$ respectively.

The optical and electrical parameters were taken from.[38–41] For the material of QDs, $E_g = 0.717$ eV, $\Delta_0 = 0.321$ eV, $m_c = 0.041 m_0$, $m_{hh} = 0.424 m_0$, $m_{lh} = 0.052 m_0$. For the material of the OCL, $E_g' = 0.997$ eV, $\Delta_0' = 0.240$ eV, $m_c' = 0.057 m_0$, $m_{hh}' = 0.471 m_0$, $m_{lh}' = 0.084 m_0$. The c- and v-band offsets at the QD–OCL heteroboundary calculated from the equations $\Delta E_{c1}/\Delta E_{g1} = 0.39$, $\Delta E_{v1}/\Delta E_{g1} = 0.61$ (see Ref.[41]) are $\Delta E_{c1} = 0.109$ eV and $\Delta E_{v1} = 0.171$ eV.

The energies of the electron and hole ground states were calculated by replacing the three-dimensional square well of a cubic QD by the effective spherical well.[25,32] For the mean size of cubic QDs $\bar{a} = 150$ Å, $\bar{\varepsilon}_n = 0.060$ eV and $\bar{\varepsilon}_p = 0.009$ eV.

At $\hbar \omega = \bar{E}_0 = E_g + \bar{\varepsilon}_n + \bar{\varepsilon}_p$, the real parts of the dielectric constant for the materials of the OCL and cladding layers were calculated using Ref.[39]: $\epsilon = 11.112$, $\epsilon' = 10.000$.

Here we suppose a Gaussian distribution of relative QD size fluctuations.

The following quantities are shown in Fig. 8 versus the RMS of relative QD size fluctuations $\delta = (\Delta \varepsilon)_{inhom}/(q_n \bar{\varepsilon}_n + q_p \bar{\varepsilon}_p)$ at various total losses β: optimum thickness of the OCL (a), optical confinement factor at $b = b^{opt}$ (b), optimum surface density of QDs (c), and minimum threshold current density (d).

For $\beta = 10\,\mathrm{cm}^{-1}$ and overall QD size fluctuations of 10% ($\delta = 0.05$, which is rather typical of the ensembles of QDs in the actual lasers,[33,42–44])

$$j_{\mathrm{th}}^{\min} \approx 8.3\ \mathrm{A\,cm}^{-2} \qquad N_{\mathrm{S}}^{\mathrm{opt}} \approx 6.2 \times 10^{10}\ \mathrm{cm}^{-2}. \qquad (54)$$

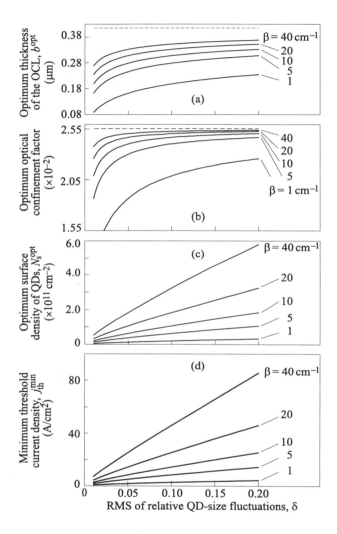

Fig. 8. Optimum thickness of the OCL b^{opt} (a), optimum optical confinement factor $\Gamma(b^{\mathrm{opt}})$ (b), optimum surface density of QDs $N_{\mathrm{S}}^{\mathrm{opt}}$ (c) and minimum threshold current density $j_{\mathrm{th}}^{\min}$ (d) against RMS of relative QD size fluctuations δ at various total losses in the waveguide β for a $\mathrm{Ga}_x\mathrm{In}_{1-x}\mathrm{As}_y\mathrm{P}_{1-y}/\mathrm{InP}$ heterostructure lasing near 1.55 μm. In (a) and (b), the horizontal dashed lines depict the b^{opt} and $\Gamma(b^{\mathrm{opt}})$ asymptotes, respectively.

The ratio of the stimulated transition rate in QDs at the lasing threshold to the spontaneous transition rate in the OCL at the transparency threshold $s \approx 4.1$.

The wavelength at the maximum gain, the optimum values of the OCL thickness and optical confinement factor are as follows: $\bar\lambda_0 = 1.580\,\mu\text{m}$, $b^{\text{opt}} = 0.279\,\mu\text{m}$, $\Gamma(b^{\text{opt}}) = 0.024$. The mean level occupancies in QDs are $\langle f_{n,p}\rangle^{\text{opt}} = 0.668$. The separation between neighboring QD centers in the optimized structure is $d^{\text{opt}} \approx 2.68\,\bar{a} = 402\,\text{Å}$.

Recall that no account has been taken of the nonradiative recombination processes in QDs and in the OCL. The inclusion of the nonradiative recombination processes in the OCL will increase somewhat j_{th}. It would appear reasonable that, for low values of j_{th}, the corrections to it caused by the Auger processes should be small. With the method of estimation of the dependence of carrier lifetime in the OCL on carrier density including all of the possible nonradiative processes,[22] we obtain the threshold current density for the structure in question to increase from the above value of 8.3 A cm^{-2} to 9.5 A cm^{-2}.

It is worth noting that for the QD size $\bar{a}$ and potential well depths $\Delta E_{c,v}$ considered here, there are excited hole states and there are no such electron states in QDs. In highly symmetrical (e.g., cubic) QDs, transitions from the ground electron level to the higher hole levels are forbidden by the selection rule. Consequently, the presence of these latter leads merely to the change in the relation between the occupancies of the ground electron and hole levels. This will have no effect on the minimum threshold current density which was shown to be independent of this relation.

4.4. Nonequilibrium filling of QDs

Inasmuch as the equation the modal gain spectrum peak versus $f_{n,p}$ is identical with (22), here too, the level occupancies in a QD will be given by (37) (in this section, we shall restrict our consideration to the specific case of $f_n = f_p$). Substituting this latter into (27) gives

$$j_{\text{th}} = Z_{\text{L}} \frac{1}{4} \frac{eN_S^{\min}}{\tau_{\text{QD}}} \left(1 + \frac{N_S^{\min}}{N_S}\right)^2 \frac{N_S}{N_S^{\min}} + \frac{1}{4} \frac{ebB}{\sigma_n \sigma_p v_n v_p \tau_{\text{QD}}^2} \frac{(1+N_S^{\min}/N_S)^4}{(1-N_S^{\min}/N_S)^2} \quad (55)$$

where $N_S > N_S^{\min}$. All the aforesaid, regarding the character of the dependence of j_{th} on structure parameters (Figs. 5 and 6), remains valid in this case as well.

A comparison of (38) and (55) shows that, by analogy with the dimensionless parameter s [see (45)], we can introduce the dimensionless parameter s' as

$$s' = \frac{(4/\xi)(\sqrt{\epsilon}/\bar\lambda_0)^2 \,((\Delta\varepsilon)_{\text{inhom}}/\hbar)\,\beta}{B/(4\sigma_n\sigma_p v_n v_p \tau_{\text{QD}}^2)}. \quad (56)$$

Here, too, s' represents the ratio of the stimulated transition rate in QDs at the lasing threshold to the spontaneous transition rate in the OCL at the transparency threshold. By analogy with s, s' plays the role of a universal dimensionless parameter controlling the magnitudes of the minimum threshold current density, optimum surface density of QDs and optimum thickness of the OCL.

In the case corresponding to region A in Fig. 2, the product of the free carrier densities at the transparency threshold, $np = n_1 p_1$, controlled by thermally excited escapes from QDs appears in (45), whereas the product of the free carrier densities at the transparency threshold, $1/(4\sigma_n\sigma_p v_n v_p \tau_{QD}^2)$, controlled by the radiative recombination in QDs appears in (56).

Clearly the equations for j_{th}^{min} and for the optimum parameters of the structure to be derived by minimizing (55), will be analogous to those corresponding to region A in Fig. 2 if the product of the free carrier densities at the transparency threshold $n_1 p_1$ in these latter is replaced by $1/(4\sigma_n\sigma_p v_n v_p \tau_{QD}^2)$. Thus, in the limiting case of $s' \gg 1$, equations identical with (43) and (50) are obtained. This result is reasonably apparent: in the limiting cases of $s' \gg 1$ and $s \gg 1$, the currents of recombination in the OCL, i.e., the second terms in (38) and (55), are negligible, whereas the first terms, i.e., the recombination currents in QDs, are identical.

As in equilibrium case, for small QD size fluctuations and/or small losses ($s' \ll 1$), the optimum parameters b^{opt}, $\Gamma(b^{opt})$, N_S^{opt}, and the minimum threshold current density j_{th}^{min} are proportional to $\sqrt{(\Delta\varepsilon)_{inhom}}$ and $\sqrt{\beta}$:

$$N_S^{opt} = \nu_2 \frac{1}{\pi} \left[\frac{1}{\xi} \frac{\epsilon}{\epsilon - \epsilon'} \frac{B}{\sigma_n \sigma_p v_n v_p} \beta \frac{(\Delta\varepsilon)_{inhom}}{\hbar} \right]^{1/2} \quad (57)$$

$$j_{th}^{min} = \nu_3 \frac{1}{2\pi} \frac{e}{\tau_{QD}} \left[\frac{1}{\xi} \frac{\epsilon}{\epsilon - \epsilon'} \frac{B}{\sigma_n \sigma_p v_n v_p} \beta \frac{(\Delta\varepsilon)_{inhom}}{\hbar} \right]^{1/2}. \quad (58)$$

The numerical constants ν_2 and ν_3 are as follows: $\nu_2 = \sqrt{3}\sqrt{5+\sqrt{13}} \approx 5.081$ and $\nu_3 = \nu_2[(5+\sqrt{13})/(4+\sqrt{13})]^2 \approx 6.505$.

Within numerical factors of the order of unity

$$\frac{(j_{th}^{min})_C}{(j_{th}^{min})_A} \propto \frac{(N_S^{opt})_C}{(N_S^{opt})_A} \propto \frac{\sqrt{(\tau_n^{esc}\tau_p^{esc})_A}}{\tau_{QD}} \ll 1 \quad (59)$$

where the subscript A stands for the quantities given by (51) and (52) and is related to the case of equilibrium filling of QDs and a narrow line of the quantized energy distribution (region A in Fig. 2), the subscript C stands for the quantities given by (57) and (58) and is related to the case of nonequilibrium filling (region C in Fig. 2). The factor $\sqrt{(\tau_n^{esc}\tau_p^{esc})_A}/\tau_{QD}$ in (59) is much less than unity. This is because the times of thermally excited escape of carriers from a QD (in the case of equilibrium filling) $(\tau_n^{esc})_A$ and $(\tau_p^{esc})_A$ are small as compared to the radiative lifetime in a QD, τ_{QD} [see (11)]. (Recall that it is precisely the magnitude of the parameter $\sqrt{(\tau_n^{esc}\tau_p^{esc})_A}/\tau_{QD}$ that determines whether the filling of QDs is equilibrium or not). Inequalities (59) are readily apparent from the following: at relatively low T, when the QD filling is nonequilibrium, the thermal escape of carriers from QDs is suppressed.

In the nonequilibrium filling of QDs, j_{th} is essentially temperature-independent. This fact owes to the temperature-independent free carrier densities in the OCL [see

(26)]. (More precisely, there is a weak T–dependence of the free carrier densities due to such dependence of the cross sections of carrier capture into a QD, $\sigma_{n,p}$).

5. Violation of the Charge Neutrality in QDs

The second [in addition to (35)] equation relating f_n to f_p (throughout Sections 5 and 6 we are dealing with the level occupancies in the mean-sized QD) and required to calculate the recombination current density of a laser should be derived from the solution of the self-consistent problem for the electrostatic field distribution across the junction. It should depend on the laser design, i. e. on the spatial distribution of donor and acceptor impurities, c- and v-band offsets, and the QD layer location. Let this second equation for f_n and f_p be written in the general case as

$$f_p - f_n = \Delta\left(\frac{N_S}{N_S^{\min}}\right) \tag{60}$$

where the specific type of the function $\Delta\left(N_S/N_S^{\min}\right)$ depends on the type of the QD laser structure. For the laser with a QD layer in the i-region of p–i–n double heterojunction, we will find Δ in the next Section.

We consider both positive and negative values of the modal gain g with an eye to treatment of the gain–current dependence of the laser (Section 5.3). The positive (negative) g corresponds to the mode of operation which is above (below) the transparency threshold. Hence, we rewrite (35) as

$$f_n + f_p - 1 = \frac{N_S^{\min}}{N_S}\operatorname{sgn}(g). \tag{61}$$

Here $N_S^{\min}$ is given by (32) where β is replaced by $|g|$.

If $g = \beta$, where β is the total loss coefficient, eq. (61) represents the lasing threshold condition (hereafter referred to as the threshold condition). If $g = 0$ ($N_S^{\min} = 0$), eq. (61) represents the transparency threshold condition.

For $g > 0$, $N_S^{\min}$ is the minimum surface density of QDs required to attain the peak modal gain value g. At $N_S = N_S^{\min}$, we have $f_{n,p} = 1$. For $g < 0$, $N_S^{\min}$ is the minimum surface density of QDs required for the peak modal absorption value of $-g$. At $N_S = N_S^{\min}$, we have $f_{n,p} = 0$.

With (60) and (61), f_n and f_p may be written as

$$f_{n,p} = \frac{1}{2} + \frac{1}{2}\frac{N_S^{\min}}{N_S}\operatorname{sgn}(g) \mp \frac{1}{2}\Delta\left(\frac{N_S}{N_S^{\min}}\right). \tag{62}$$

Hereafter the upper and lower signs correspond to "n" and "p" subscripts respectively.

With (62), the spontaneous recombination current density may be written as

$$j = \left(\frac{eN_S^{\min}}{\tau_{QD}} + ebBn_1p_1\right) + \frac{eN_S^{\min}}{\tau_{QD}}W + ebBn_1p_1\frac{1}{W} \tag{63}$$

above the transparency threshold and as

$$j = \frac{eN_S^{\min}}{\tau_{QD}} W + ebBn_1 p_1 \frac{1}{1 + \frac{1}{W}} \quad (64)$$

below the transparency threshold.

The function $W\left(N_S/N_S^{\min}\right)$, where $N_S > N_S^{\min}$, is

$$W\left(\frac{N_S}{N_S^{\min}}\right) = \frac{1}{4} \frac{N_S}{N_S^{\min}} \left[\left(1 - \frac{N_S^{\min}}{N_S}\right)^2 - \Delta^2\left(\frac{N_S}{N_S^{\min}}\right)\right]. \quad (65)$$

In the general case the function $\Delta\left(N_S/N_S^{\min}\right)$ must satisfy the following conditions:

$$\Delta(1) = 0 \qquad \Delta(\infty) = 0. \quad (66)$$

The first condition is evident from (61): when $N_S = N_S^{\min}$, $f_n = f_p = 1$ for $g > 0$, and $f_n = f_p = 0$ for $g < 0$. As for the second condition, we will give proof to it in the next Section; for now, we note that in the limiting case of $N_S \to \infty$, a QD layer turns into a QW wherein the charge neutrality holds.

As may be seen from (65) and (66), W increases monotonically from zero to infinity as N_S increases from $N_S^{\min}$ to infinity. Thus, it is easy to see from (63) and (65) that in the general case of neutrality violation the character of the dependence of j on N_S for the "above transparency threshold" operating mode is identical to that for the specific case of symmetric structure considered in Section 4.3.1 [Fig. 5(a)]. This dependence is nonmonotonic as before: if $N_S \to N_S^{\min}$ or if $N_S \to \infty$, j increases infinitely (Fig. 15). The dependence of j on N_S has a minimum at some one value of N_S. For the lasing threshold, this optimum value of N_S will be calculated in Section 5.2.

For the "below transparency threshold" operating mode, j increases monotonically from zero to infinity as N_S increases from $N_S^{\min}$ to infinity.

For $N_S \gg N_S^{\min}$, j for both above and below transparency threshold operating modes tends to the value $j_{tr} = (1/4)\,(eN_S/\tau_{QD}) + ebBn_1 p_1$ which is the transparency current density for large N_S [the inclined dashed line in Fig. 15].

In the case of arbitrary N_S, the following equation for j_{tr} may be obtained from (20) using the transparency threshold condition $f_n + f_p - 1 = 0$:

$$j_{tr} = \frac{eN_S}{\tau_{QD}} f_n(1 - f_n) + ebBn_1 p_1 = \frac{1}{4} \frac{eN_S}{\tau_{QD}} \left[1 - \Delta^2(N_S)\right] + ebBn_1 p_1. \quad (67)$$

5.1. Calculations for a p–i–n heterostructure

The cladding layers are p- and n-sides, and the OCL is the i-region of the structure (Fig. 9). To minimize the threshold current, the QD layer is located in the center of the OCL. When solving the problem for the electrostatic field distribution across the junction, a QD layer charge, governing the step in the field at the layer, has been properly taken into account.

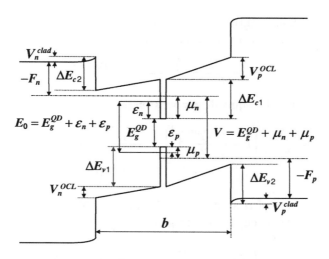

Fig. 9. Energy band diagram of a QD laser structure. The QD layer is considered as the charged sheet. The static electric field associated with the built-in potential and with the applied bias is uniform in each of two parts of the i-region. Since the QD layer is charged, the electric field differs at the left- and right-hand sides of the OCL.

The following equations have been obtained for $f_{n,p}$ and $\Delta = f_p - f_n$:

$$f_{n,p}(\aleph_{as}, \alpha, \phi) = [\exp(-\phi \mp v) + 1]^{-1} \qquad (68)$$

$$\Delta(\aleph_{as}, \alpha, \phi) = \frac{v + \aleph_{as}}{\alpha} \qquad (69)$$

where the dimensionless quantity v is the solution of the algebraic equation

$$-v - \aleph_{as} = \alpha \frac{\sinh v}{\cosh \phi + \cosh v}. \qquad (70)$$

Here the dimensionless parameter of asymmetry $\aleph_{as}$ and the parameter α are

$$\aleph_{as} = \frac{(\Delta E_v + F_p - \varepsilon_p) - (\Delta E_c + F_n - \varepsilon_n)}{2T} \qquad (71)$$

$$\alpha = \frac{e^2 N_S/(4T)}{\epsilon_s/(4\pi b)} = \frac{(\pi/\epsilon_s) e^2 b N_S}{T} \qquad (72)$$

where $\Delta E_{c,v} = \Delta E_{c1,v1} + \Delta E_{c2,v2}$, $\Delta E_{c1,v1}$ and $\Delta E_{c2,v2}$ are the conduction and valence band offsets at the QD–OCL and at the OCL–cladding heteroboundaries, respectively, $F_{n,p} = -T \ln \left(N_{c,v}^{clad}/N_{D,A} \right)$ are the quasi-Fermi levels in n- and p-claddings respectively away from the heteroboundaries with the OCL (Fig. 9), $N_{c,v}^{clad}$ are the conduction and valence band effective densities of states for the material of the claddings, N_D (N_A) is the donor (acceptor) impurity concentration in n-cladding (p-cladding), and ϵ_s is the static dielectric constant.

The dimensionless quantity ϕ is simply the applied bias V reckoned from the transition energy in the mean-sized QD E_0 and normalized to $2T$:

$$\phi = \frac{V - E_0}{2T} = \frac{(\mu_n - \varepsilon_n) + (\mu_p - \varepsilon_p)}{2T}. \tag{73}$$

Equation (70) may be equivalently written as

$$-v - \aleph_{as} = \alpha_m \frac{N_S}{N_S^{min}} \left[1 - \left(\frac{N_S^{min}}{N_S}\right)^2\right] \frac{\sinh v}{\cosh v + \sqrt{1 + \left(\frac{N_S^{min}}{N_S}\right)^2 \sinh^2 v}} \tag{74}$$

where α_m is the value of α at $N_S = N_S^{min}$. In (74), N_S^{min} is governed by g. The electron and hole level occupancies are given by (62) where

$$\Delta\left(\aleph_{as}, \alpha_m, \frac{N_S}{N_S^{min}}\right) = \frac{v + \aleph_{as}}{\alpha_m} \frac{N_S^{min}}{N_S}. \tag{75}$$

Equation (75), wherein v is the solution of (74) [or, alternatively, eq. (69), wherein v is the solution of (70)] is the second [in addition to (61)] equation relating f_n to f_p. When coupled with (61), eq. (75) [or (69)] makes it possible to calculate $f_{n,p}$.

Equations (68)–(70) will be exploited to analyze the current–voltage characteristic of the laser (Section 5.4). To calculate the dependences on the surface density of QDs, we make use of (74) and (75).

It is the dimensionless parameter α_m (or α) and the parameter of asymmetry $\aleph_{as}$ that are of first importance and that govern the magnitude of $\Delta = f_p - f_n$. As will be seen from (74) and (75), $v(-\aleph_{as}, \alpha_m, N_S^{min}/N_S) = -v(\aleph_{as}, \alpha_m, N_S^{min}/N_S)$ and $\Delta(-\aleph_{as}, \alpha_m, N_S^{min}/N_S) = -\Delta(\aleph_{as}, \alpha_m, N_S^{min}/N_S)$. For this reason, we restrict the discussion to the case of positive parameter of asymmetry $\aleph_{as}$. It will be also seen from thence that $-\aleph_{as} < v < 0$ and $\Delta > 0$ for positive $\aleph_{as}$.

At $\aleph_{as} = 0$, $v = 0$ and $\Delta = 0$ no matter what the surface density of QDs N_S and parameter α_m. Hence the criterion for the charge neutrality in QDs at any N_S is

$$\Delta E_c + F_n - \varepsilon_n = \Delta E_v + F_p - \varepsilon_p. \tag{76}$$

Criterion for the charge neutrality depends on the total band offsets between the QDs and claddings $\Delta E_{c,v} = \Delta E_{c1,v1} + \Delta E_{c2,v2}$, quantized energy level positions $\varepsilon_{n,p}$ (which in turn depend on the QD size and band offsets between the QDs and the OCL), and the quasi-Fermi level positions in n- and p-claddings $F_{n,p}$ (i. e. on the doping levels of claddings). In Sections 4.3.1 and 4.3.2, the charge neutrality was assumed, which implies that the doping levels of claddings were matched to satisfy criterion (76).

Since the quasi-Fermi levels in the claddings, $F_{n,p}$, depend on the temperature, criterion for the charge neutrality also depends on T. This means that the QD layer may be neutral only at one certain T [at which (76) holds]; at any other T it should be charged. Throughout this section, the results for room temperature operation of a laser are presented.

The greater $\aleph_{as}$, the greater $\Delta = f_p - f_n$ at given N_S/N_S^{min} and α_m (Fig. 10). The dashed curve in the figure depicts Δ for $\aleph_{as} \to \infty$. It is given by the asymptotic equation $\Delta = 1 - N_S^{min}/N_S$ which is readily obtainable from (74) and (75). With (62), $f_n = 0$ and $f_p = 1 - (N_S^{min}/N_S)$ for $g < 0$. For $g > 0$, $f_n = N_S^{min}/N_S$ and $f_p = 1$. Thus, in the limiting case of large positive $\aleph_{as}$, for the "below transparency threshold" operating mode, f_n is pinned at its minimum value (i. e., the electron level is fully unoccupied), whereas f_p changes with N_S. For the "above transparency threshold" operating mode, f_p is pinned at its maximum value (i. e., the hole level is fully occupied), whereas f_n changes with N_S.

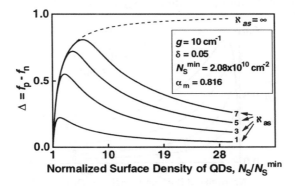

Fig. 10. Difference of the hole and electron level occupancies in QDs against normalized surface density of QDs at various values of the parameter of asymmetry $\aleph_{as}$.

All the results and deductions made above and to be made below for $\aleph_{as} > 0$ will remain valid for $\aleph_{as} < 0$ if we interchange f_n and f_p.

We now turn to the parameter α_m. The differential capacitance (per unit area) of a QD layer, $C_{dif} = e^2 N_S |\partial(f_p - f_n)/\partial V|$, may be written as

$$C_{dif} = \frac{1}{2} \frac{e^2 N_S}{T} \frac{|(f_p - f_n)(f_n + f_p - 1)|}{1 + (\pi e^2 b N_S/\epsilon_s T)\left[(f_p - f_p^2) + (f_n - f_n^2)\right]} \quad (77)$$

or, using (61), as

$$C_{dif} = \frac{1}{2} \frac{e^2 N_S^{min}}{T} \frac{|f_p - f_n|}{1 + (\pi e^2 b N_S/\epsilon_s T)\left[(f_p - f_p^2) + (f_n - f_n^2)\right]}. \quad (78)$$

Equation (77), where $f_{n,p}$ are given by (68) represents the capacitance–voltage dependence of the QD layer.

Thus, α_m is the ratio of the peak differential capacitance of the QD layer, $e^2 N_S^{min}/(4T)$, to that of the OCL, $\epsilon_s/(4\pi b)$. The smaller α_m, the greater $\Delta = f_p - f_n$ at given N_S/N_S^{min} and $\aleph_{as}$. The difference of f_p and f_n should fall off with increasing capacitance of a QD layer. It is the magnitude of the effective capacitance that is responsible for the drastic difference between a QD layer and a QW with respect to the charge neutrality. Owing to the high effective capacitance of a QW the neutrality violation in a QW is hampered. In QD lasers, contrastingly, at actual values of N_S required for lasing to occur, the effective capacitance of a QD layer is not high enough for the difference of f_p and f_n to be suppressed. This is why, in contrast to a QW, the QD layer is charged and $f_n \neq f_p$.

It is significant that the smaller the QD size dispersion, i. e. the more perfect the laser structure, and/or the smaller are the total losses, the smaller α_m [see eq. (32) for $N_S^{\min}$] and hence the greater Δ. Shown in Fig. 11 is a set of curves for Δ versus $N_S/N_S^{\min}$ at various values of RMS of relative QD size fluctuations (also given are the corresponding values of α_m). The $\aleph_{as}$ value is 2.263 which corresponds to the equal doping levels of n- and p-claddings ($N_D = N_A$) for the structure considered.

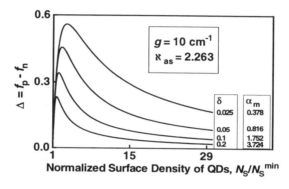

Fig. 11. Difference of the hole and electron level occupancies in QDs against normalized surface density of QDs at various values of the RMS of relative QD size fluctuations δ.

As will be seen from Figs. 10 and 11 [and from (74) and (75)], $\Delta = 0$ at $N_S = N_S^{\min}$ which has a simple explanation. For the "above transparency threshold" operating mode, both the electron and hole levels are fully occupied ($f_n = f_p = 1$) at $N_S = N_S^{\min}$. For the "below transparency threshold" operating mode, both the electron and hole levels are fully unoccupied ($f_n = f_p = 0$) at $N_S = N_S^{\min}$. It can also be seen from the figures that Δ goes to zero as $N_S/N_S^{\min} \to \infty$. The reason is that if the surface density of QDs tends to infinity, a layer with QDs turns to a QW for which the charge neutrality holds.

The difference of f_p and f_n peaks at one value of N_S [see eqs. (33)-(35) in Ref.[26]]. For $\alpha_m = 0.816$ and $\aleph_{as} = 2.263$ ($N_D = N_A$), the peak value $(f_p - f_n)^{\max} = 0.456$ is obtained at $N_S \approx 2.58 N_S^{\min} \approx 5.36 \times 10^{10}\,\text{cm}^{-2}$. For these values of α_m and $\aleph_{as}$, Fig. 12 depicts $f_{n,p}$ versus $N_S/N_S^{\min}$ for below (a) and above (b) transparency threshold operating modes.

Hence our results reveal that if $\aleph_{as} \neq 0$, the charge neutrality is broken down in QDs unless $N_S = N_S^{\min}$ (f_n and f_p are both zero or unity together in "below transparency threshold" or "above transparency threshold" operating modes respectively) or $N_S/N_S^{\min} \to \infty$ (f_n and f_p are both 1/2 together).

It follows from (70) and (68) that $f_n(-\phi) = 1 - f_p(\phi)$. The dependences of $f_{n,p}$ and $\Delta = f_p - f_n$ on the normalized applied bias $\phi = (V - E_0)/2T$ are shown in Figs. 13(a) and (b). The applied voltage in eV is also indicated on the abscissa (the top axis). We notice that $f_p - f_n$ is at its maximum at $\phi = 0$, that is at the applied bias value $V = E_0 = E_g + \varepsilon_n + \varepsilon_p$ corresponding to the transparency threshold. This is because the differential capacitance of the QD layer is a minimum, namely zero, at the transparency threshold [see (77) and (78)].

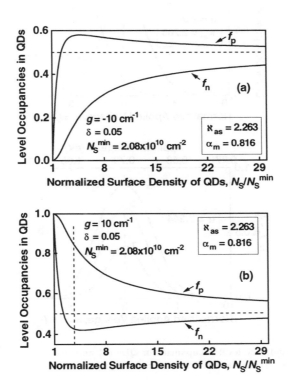

Fig. 12. Electron and hole level occupancies in QDs against normalized surface density of QDs for below (a) and above (b) transparency threshold operating modes. In (b), the intersections of the vertical dashed line and the solid curves depict the values of f_n and f_p which correspond to the optimum surface density of QDs [given by (80)] minimizing the threshold current density.

The peak modal gain and the population inversion in QDs are zero at the transparency threshold: $g = 0$ and $f_n + f_p - 1 = 0$. (If the charge neutrality in QDs were the case, this would give immediately $f_{n,p} = 1/2$.) Inserting $\phi = 0$ into (70), (68), and (69) gives

$$f_{n,p} = \frac{1}{2} \mp \frac{\Delta}{2} \qquad \Delta = -\tanh\frac{v}{2} \qquad (79)$$

where v is the solution of the equation $-v - \aleph_{as} = \alpha \tanh(v/2)$. In the limiting case of small N_S ($\alpha \ll \aleph_{as}$), $v = -\aleph_{as}$ and f_n and f_p differ significantly from each other: $f_{n,p} = 1/2 \mp (1/2)\tanh(\aleph_{as}/2)$, $\Delta = \tanh(\aleph_{as}/2)$. In the opposite limiting case of large N_S ($\alpha \gg \aleph_{as}$), $v = -2\aleph_{as}/\alpha$ and f_n and f_p are close to each other: $f_{n,p} = 1/2 \mp (1/2)(\aleph_{as}/\alpha)$, $\Delta = \aleph_{as}/\alpha$. Fig. 14 shows $f_{n,p}$ and $\Delta = f_p - f_n$ versus N_S (contrast with Fig. 12).

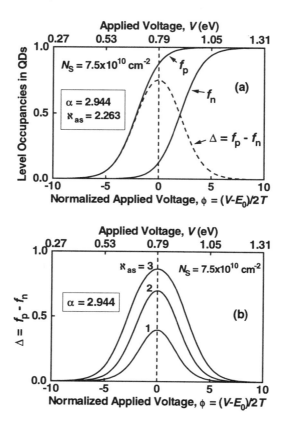

Fig. 13. Electron and hole level occupancies in QDs (a) and the difference of them (b) against applied voltage. The vertical dashed lines mark the transparency threshold.

As seen from (62) and (68)–(75), the difference of the hole and electron level occupancies and hence the level occupancies themselves are temperature-dependent (note that parameters $\aleph_{as}$, α, α_m, and ϕ depend on T). This strongly effects the T-dependence of the threshold current. We will turn to this issue in Section 6.

5.2. Effect of neutrality violation on the optimization of a laser

Self-consistent consideration of the QD charge is of first importance for the optimization of a laser. Inserting $f_{n,p}$ into (47) gives the following equation for N_S^{opt}:

$$\frac{N_S^{opt}}{N_S^{min}} = 1 + 2\left(\frac{b\Gamma}{a}\frac{1}{s}\right)^{1/2} + \left\{\left[1 + 2\left(\frac{b\Gamma}{a}\frac{1}{s}\right)^{1/2}\right]^2 - 1 + \left(\frac{v^{opt} + \aleph_{as}}{\alpha_m}\right)^2\right\}^{1/2} \quad (80)$$

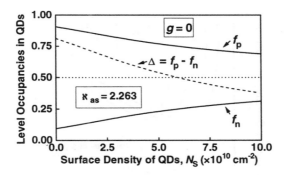

Fig. 14. Electron and hole level occupancies in QDs (the solid curves) and the difference of them (the dashed curve) against surface density of QDs at the transparency threshold. The horizontal dotted line depicts $f_{n,p}$ if the charge neutrality were the case (if $\aleph_{as} = 0$).

where v^{opt} is to be found from (74) wherein the ratio N_S/N_S^{min} is replaced by its optimum value N_S^{opt}/N_S^{min} given by (80).

A comparison of (80) and (49) shows that the inclusion of the fact of charge neutrality violation gives rise to the term $[(v^{opt} + \aleph_{as})/\alpha_m]^2$ in the equation for N_S^{opt}. This term increases with increasing $\aleph_{as}$; it is zero when the charge neutrality holds ($\aleph_{as} = 0$). Consequently, N_S^{opt} increases with increasing charge neutrality violation in QDs. This is illustrated in Fig. 15 where j_{th} is plotted versus N_S/N_S^{min} at various values of $\aleph_{as}$. The inset shows the relative difference of N_S^{opt} given by (80) and (49) versus the parameter of asymmetry $\aleph_{as}$.

For $\alpha_m = 0.816$ [which corresponds to the overall QD size fluctuations of 10% ($\delta = 0.05$) and to the total losses $\beta = 10 \text{ cm}^{-1}$] and $\aleph_{as} = 2.263$, $N_S^{opt} \approx 3.70 N_S^{min} \approx 7.69 \times 10^{10} \text{ cm}^{-2}$. That one calculated assuming the charge neutrality is $6.20 \times 10^{10} \text{ cm}^{-2}$ [see (54)]. At $N_S = N_S^{opt}$, the electron and hole level occupancies and difference of them are as follows [see Fig. 12(b)]: $f_n^{opt} = 0.426$ [$(\mu_n - \varepsilon_n)/T = -0.30$], $f_p^{opt} = 0.845$ [$(\mu_p - \varepsilon_p)/T = 1.69$], and $(f_p - f_n)^{opt} = 0.419$; $(f_p - f_n)^{opt}$ is thus seen to be close to the maximum value of $f_p - f_n$ (see above).

For the optimized structure, the actual and normalized applied voltages at the lasing threshold are $V = 0.822 \text{ eV}$ and $\phi = 0.697$. For the equal doping levels of claddings $N_D = N_A = 5 \times 10^{17} \text{ cm}^{-2}$, the built-in voltage is $V_b = 1.274 \text{ eV}$. The potential energy change and the electrostatic field in the left-hand part of the i-region are $V_n^{OCL} = 0.193 \text{ eV}$ and $\mathcal{E}_n = 1.384 \times 10^4 \text{ V/cm}$ (Fig. 9). Those in the right-hand part are $V_p^{OCL} = 0.259 \text{ eV}$ and $\mathcal{E}_p = 1.856 \times 10^4 \text{ V/cm}$.

As mentioned in Section 2, the analysis presented is carried out in the framework of the mean field approximation. The approximation neglects the electron–electron and electron–hole interaction within a QD. Under the optimum conditions, when j_{th} is a minimum, the average number of electrons per one QD is less than unity ($f_n^{opt} = 0.426$). For this reason, electron–electron interaction within a QD is of no importance. As for the electron–hole interaction within a QD, it leads to the red

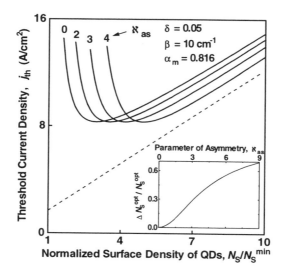

Fig. 15. Threshold current density against normalized surface density of QDs at various values of the parameter of asymmetry $\aleph_{as}$. The inclined dashed line depicts the transparency current density at $N_S \to \infty$. The inset shows the relative difference of the actual N_S^{opt} [given by (80)] and that calculated using $\aleph_{as} = 0$ [given by (49)].

shift of the transition energy in QDs and to some enhancement of the matrix element of the dipole moment of optical transitions in QDs. Mean field approximation also neglects correlation corrections due to the interaction of carriers in one QD with carriers in other QDs and with carriers in the continuous spectrum states. The carrier correlation effects should result in additional broadening of the lasing line. The characteristic energy of this broadening should be of the order of $e^2/(\epsilon_s d)$, where $d \approx 1/\sqrt{N_S}$ is the average distance between the adjacent QDs. In view of the fact that the average two-dimensional threshold carrier densities for QD lasers are at least one order of magnitude less than those for QW lasers, d is high enough and the broadening energy is small compared to inhomogeneous line broadening caused by fluctuations in QD sizes. Thus, at $N_S = N_S^{opt} \approx 7.69 \times 10^{10}$ cm^{-2}, $(e^2/\epsilon_s)\sqrt{N_S} =$ 3 meV which is 3 times less than inhomogeneous broadening corresponding to RMS of relative QD size fluctuations $\delta = 0.05$.

5.3. Gain–current dependence

Equation (62) may be written as follows:

$$f_{n,p} = \frac{1}{2} + \frac{1}{2}\frac{g}{g^{max}} \mp \frac{1}{2}\Delta\left(\frac{g}{g^{max}}\right). \tag{81}$$

Here $\Delta(g/g^{max})$ is given by (69), wherein $v = v(g/g^{max})$ is the solution of the equation

$$-v - \aleph_{as} = \alpha \left[1 - \left(\frac{g}{g^{max}}\right)^2\right] \frac{\sinh v}{\cosh v + \sqrt{1 + \left(\frac{g}{g^{max}}\right)^2 \sinh^2 v}}. \quad (82)$$

It follows from (81) and (82) that $f_n(-g) = 1 - f_p(g)$. Fig. 16 shows a set of curves for Δ versus g/g^{max} at various values of $\aleph_{as}$. Δ is zero both when $g = -g^{max}$ (when $f_{n,p} = 0$) and $g = g^{max}$ (when $f_{n,p} = 1$). As already mentioned, Δ peaks at the transparency threshold $g = 0$.

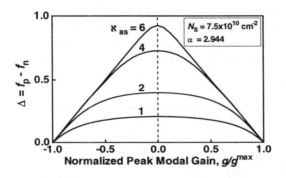

Fig. 16. Difference of the hole and electron level occupancies in QDs against normalized peak modal gain at various values of $\aleph_{as}$. The vertical dashed line marks the transparency threshold.

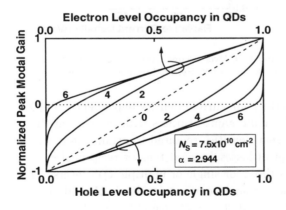

Fig. 17. Normalized peak modal gain against electron and hole level occupancies in QDs at various values of $\aleph_{as}$. The corresponding values of $\aleph_{as}$ are given adjacent to the curves. The dashed line corresponds to the case of $\aleph_{as} = 0$.

Fig. 17 shows a set of curves for the normalized peak modal gain g/g^{max} versus f_p (the bottom x-axis) and f_n (the top x-axis) at various values of $\aleph_{as}$. These curves for QD lasers are the analogue of the gain–carrier density curves for QW or bulk lasers. The dashed line depicts $g/g^{\max}$ versus f_p (and f_n) if the charge neutrality in QDs were the case. As evident from the figure, non-linearity of the curves is caused by the neutrality violation.

Substituting (81) into (20) gives

$$j(g) = \frac{1}{4} \frac{eN_S}{\tau_{QD}} \left[\left(1 + \frac{g}{g^{\max}}\right)^2 - \Delta^2 \left(\frac{g}{g^{\max}}\right) \right] + ebBn_1p_1 \frac{\left(1 + \frac{g}{g^{\max}}\right)^2 - \Delta^2 \left(\frac{g}{g^{\max}}\right)}{\left(1 - \frac{g}{g^{\max}}\right)^2 - \Delta^2 \left(\frac{g}{g^{\max}}\right)}. \tag{83}$$

Equation (83) where $\Delta(g/g^{\max})$ is given by (69) and (82) presents the spontaneous radiative recombination current density as a function of the peak modal gain. The inverse function presents the gain–current density dependence of the laser. If $g = \beta$, eq. (83) gives the threshold current density of the laser.

Fig. 18 shows the curves for $f_{n,p}$ versus the current density. These curves for QD lasers are the analogue of the carrier density–current density curves for QW or bulk lasers.

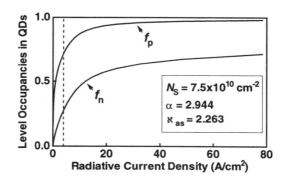

Fig. 18. Electron and hole level occupancies in QDs against radiative current density. The vertical dashed line marks the transparency threshold at which $f_p - f_n$ peaks.

Fig. 19 shows the gain–current density dependence of QD laser at various values of N_S. The peak modal gain saturates with increase in the current density. Nevertheless, the current densities required to attain the peak modal gain values that overcome the typical losses (several tens of reciprocal centimeters), are considerably smaller than those for single or multiple QW lasers. This is also evident from Fig. 20 where the threshold current density is plotted versus the total losses β at various values of N_S. At $\beta = g^{\max}$ [see (13) for $g^{\max}$], the threshold current tends to infinity. For $\beta > g^{\max}$, lasing cannot be attained at a given surface density of QDs N_S. An increase in N_S is required to attain lasing for such losses.

5.4. Current–voltage dependence

Current flowing across the junction is the difference of the radiative recombination and generation currents. The radiative recombination current density is given by

(20). The generation current density j_g may also be obtained from (20) on putting the voltage $V = 0$ there (that is, on putting $\phi = -E_0/2T$). Thus, we obtain the following equation for the current density–voltage characteristic of a QD laser:

$$j = \frac{eN_S}{\tau_{QD}}(f_n f_p - f_n f_p|_{V=0}) + ebBn_i^2\left[\exp\left(\frac{V}{T}\right) - 1\right] \tag{84}$$

where $n_i = \sqrt{N_c N_v}\exp\left(-E'_g/2T\right)$ is the intrinsic carrier density in the OCL. The voltage dependence of $f_{n,p}$ is given by (70) and (68); this is plotted in Fig. 13.

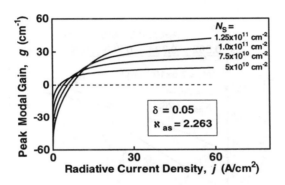

Fig. 19. Gain–current density dependence of a QD laser at various values of N_S.

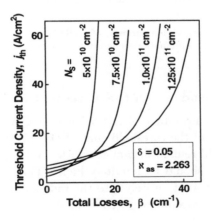

Fig. 20. Threshold current density against total losses β at various values of N_S.

Notice that the voltage dependence of the second term in (84), presenting the recombination–generation current density in the OCL, is typical of the bimolecular (quadratic in the carrier density) recombination.

The electron and hole level occupancies are practically zero at $V = 0$. Correspondingly, practically negligible is the generation current density j_g ($-j_g$ represents the saturation value of the reverse current density at $V \to -\infty$). For the specific structure considered, $f_n|_{V=0} = 6.12 \times 10^{-8}$, $f_p|_{V=0} = 1.21 \times 10^{-6}$ and $j_g = 1.30 \times 10^{-12}\,\text{A/cm}^2$. For this reason, the current density–voltage characteristic is shown for direct bias only (Fig. 21). The current densities associated with the recombination–generation in QDs and OCL, j_{QD} and j_{OCL} [the first and second terms in (84) respectively], are also shown separately in the figure. Shown in the inset is j_{QD} versus the bias. As the bias applied increases from $-\infty$ to $+\infty$, j_{QD} increases from $-(eN_S/\tau_{\text{QD}})f_n f_p|_{V=0}$ (which is practically zero) to $(eN_S/\tau_{\text{QD}})(1 - f_n f_p|_{V=0}) = 16.95\,\text{A/cm}^2$ (the horizontal dashed line in the inset).

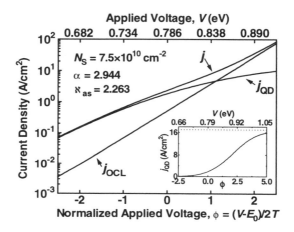

Fig. 21. Current density–voltage characteristic of a QD laser. Shown in the inset is the current density associated with the recombination–generation in QDs.

6. Temperature Dependence of the Threshold Current

An important expected advantage of QD lasers over QW lasers is extremely weak temperature sensitivity of the threshold current. Ideally, threshold current density of a QD laser j_{th} should remain unchanged with the temperature and the characteristic temperature T_0 should be infinitely high.[2] This would be so indeed if the overall injection current went entirely into the radiative recombination in QDs. In fact, because of the presence of free carriers in the OCL, a fraction of the current goes into the recombination processes therein;[22–31] this component of j_{th}, associated with the thermal escape of carriers from QDs, j_{OCL}, depends exponentially on T. Besides, violation of the charge neutrality in QDs [26] causes also the j_{th} component, associated with the radiative recombination in QDs, j_{QD}, to be temperature dependent (see below). As a result the total threshold current density, $j_{\text{th}} = j_{\text{QD}} + j_{\text{OCL}}$, should become temperature dependent,[23–31] especially at high T. Hence the characteristic temperature T_0 should become finite. This has been observed experimentally in Ref.[5] for the first time. In this section, we present a de-

6.1. Threshold current density and its components

In Section 4, for $T < T_g$, when the QD filling is nonequilibrium, j_{th} has been shown to be essentially temperature-independent. For $T > T_g$, when the QD filling is equilibrium, the temperature dependence of j_{th} has been revealed. Here we consider equilibrium filling of QDs in detail. In this case, j_{QD} and j_{OCL} are given by

$$j_{QD} = \frac{eN_S}{\tau_{QD}} f_n f_p \tag{85}$$

$$j_{OCL} = ebBn_1 p_1 \frac{f_n f_p}{(1-f_n)(1-f_p)} = ebBn_1 p_1 \frac{j_{QD}}{j_{QD} - \frac{eN_S^{min}}{\tau_{QD}}}. \tag{86}$$

In (86), the lasing threshold condition in the form of (35) is used.

Assuming the charge neutrality in QDs, $f_n = f_p$, we would obtain temperature-independent $f_{n,p}$ [see (37)]. In that case, j_{QD} would be temperature-independent. If, in addition, recombination in the OCL [see (86)] is ignored, j_{th} would be independent of T and the characteristic temperature would be infinitely high.

In fact, it is impracticable to ignore the radiative recombination in the OCL. It is this recombination channel that is primarily responsible for the temperature dependence of j_{th} of a QD laser at high T,[5,23-31] With (17) for n_1 and p_1, using $B(T) \propto T^{-3/2}$ [see (8)], and assuming the charge neutrality, we may present the T-dependence of $j_{OCL, neutral}$ as

$$j_{OCL, neutral}(T) \propto Bn_1 p_1 \propto T^{3/2} \exp\left(-\frac{\Delta E_{g1} - \varepsilon_n - \varepsilon_p}{T}\right) \tag{87}$$

where $\Delta E_{g1} = \Delta E_{c1} + \Delta E_{v1}$ is the bandgap difference between the OCL and QDs.

Correct consideration of the QD charge reveals the temperature dependence of $f_{n,p}$ (which are the analogue of the carrier densities for QW or bulk lasers):

$$f_{n,p}\left(\frac{N_S}{N_S^{min}}, T\right) = \frac{1}{2} + \frac{1}{2}\frac{N_S^{min}}{N_S} \mp \frac{1}{2}\Delta\left(\frac{N_S}{N_S^{min}}, T\right). \tag{88}$$

Here, the upper sign ("−") and lower sign ("+") correspond to "n" and "p" subscripts, respectively. The difference of the hole and electron level occupancies in QDs, $\Delta(N_S/N_S^{min}, T)$, determining the QD charge, is temperature-dependent. For the laser with a QD layer in the i-region of p–i–n junction, $\Delta(N_S/N_S^{min}, T)$ is given by (75) and (74).

Thus, violation of the charge neutrality in QDs causes the temperature dependence of j_{QD} by itself. As may be seen from (86), it also leads to the extra temperature dependence of j_{OCL} through the such dependences of $f_{n,p}$ [in addition

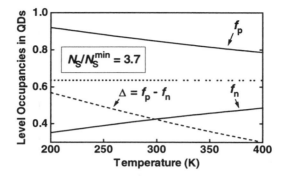

Fig. 22. Electron and hole level occupancies in QDs at the lasing threshold and the difference of them against temperature. The horizontal dotted line shows $f_{n,p}$ calculated assuming the charge neutrality in QDs. $\beta = 10\,\text{cm}^{-1}$ and $\delta = 0.05$; $N_S^{\min} = 2.1 \times 10^{10}\,\text{cm}^{-2}$. N_S and b are equal to their optimum values at $T = 300\,\text{K}$, $3.7\,N_S^{\min}$ and $0.28\,\mu\text{m}$, respectively.

to the temperature dependence of j_{OCL} through the temperature dependence of $Bn_1 p_1$, see (87)].

An examination of (88), (75), and (74) shows that the difference of the hole and electron level occupancies $\Delta\left(N_S/N_S^{\min}, T\right)$ drops slowly with T; f_n and f_p tend slowly to each other (Fig. 22). Thus, violation of the charge neutrality in QDs is suppressed with T.

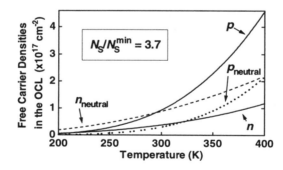

Fig. 23. Free-electron and -hole densities in the OCL at the lasing threshold against temperature. The dashed (dotted) curve depicts n (p) calculated assuming the charge neutrality in QDs.

For $\aleph_{\text{as}} > 0$ [see (71)], the actual f_n (f_p) is less (greater) than that calculated assuming the neutrality.[26] Hence it follows from (16) that the actual n (p) is less (greater) than that calculated assuming the charge neutrality (Fig. 23). (For $\aleph_{\text{as}} < 0$, f_n and f_p, as well as n and p, change places.) For this reason the product of n and p (f_n and f_p) determining j_{OCL} (j_{QD}) is less affected by the inclusion of the neutrality violation than the individual values of n and p (f_n and f_p).

The product of f_n and f_p, determining j_{QD}, is at its maximum when $f_n = f_p$, provided $f_n + f_p$ is fixed [see (35)]. For this reason j_{QD} is less than its value calculated assuming the charge neutrality (Fig. 24). From this result and from (86) it follows that j_{OCL} should be greater than its value calculated assuming the neutrality. Since f_n and f_p tend to each other with T, j_{QD} increases and tends to its value calculated assuming the neutrality (Fig. 24).

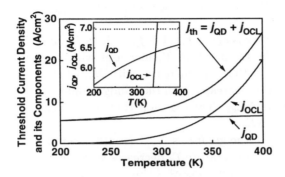

Fig. 24. Threshold current density and its components against temperature for $N_S/N_S^{min} = 3.7$. The inset shows $j_{QD}(T)$ and $j_{OCL}(T)$ on an enlarged (along the vertical axis) scale. The dashed line depicts j_{QD} calculated assuming the charge neutrality in QDs. At $T_d = 344$ K, $j_{QD} = j_{OCL}$.

The T-dependences of $f_{n,p}$ are much weaker compared to that of the product Bn_1p_1 [which is the exponential, see (87)]. Consequently, the T-dependence of j_{QD} is much weaker compared to that of j_{OCL} (Fig. 24). Nevertheless, the conclusion that j_{QD} does depend on T is of great importance. The matter is that, in the properly designed laser structures, the recombination channel in the OCL (i.e., the leakage current) should be suppressed. This should be at least one way to optimize the performance of QD lasers. *Even so, the threshold current, being determined solely by the radiative recombination in QDs, will be temperature dependent.*

The recombination current density in the OCL, j_{OCL}, increases exponentially with T, being characterized by a high activation energy $\Delta E_{g1} - \varepsilon_n - \varepsilon_p$ [see (87) and Fig. 24]. Because of this, to describe the temperature dependence of $j_{th} = j_{QD} + j_{OCL}$, we can conveniently introduce the temperature T_d at which j_{OCL} reaches j_{QD}: $j_{OCL}(T_d) = j_{QD}(T_d)$. For T fairly less than T_d, $j_{OCL} \ll j_{QD}$ and j_{th} depends only weakly on T. Conversely, for T fairly greater than T_d, $j_{OCL} \gg j_{QD}$ and j_{th} depends strongly on T (Fig. 24). With (85), (86), (8), and (17), the following equation may be derived for T_d:

$$T_d = \frac{\Delta E_{g1} - \varepsilon_n - \varepsilon_p}{\ln\left\{\frac{b N_{cv}^{red}(T_d)}{N_S[1-f_n(T_d)][1-f_p(T_d)]}r\right\}} \tag{89}$$

where $N_{cv}^{red}(T) = 2(m'_{chh}T/2\pi\hbar^2)^{3/2} + 2(m'_{clh}T/2\pi\hbar^2)^{3/2}$ is the effective reduced density of states of the conduction and valence bands for the OCL. The factor $r = \frac{1}{2}(E'_g/E_0)(P'/P)^2$ is of the order of unity.

There is certain analogy between T_d so defined and the temperature of depletion (ionization) of impurity centers. The numerator of (89) presents the energy of excitation of the electron-hole pair from a QD (an analogue of the impurity ionization energy); N_S/b is the QD number per unit volume of the OCL (an analogue of the impurity concentration).

With $N_S = N_S^{\min}$ ($f_{n,p} = 1$), (89) yields $T_d = 0$ (Fig. 30). The point is that $j_{OCL} \to \infty$ with $N_S \to N_S^{\min}$ [see (86)], whereas j_{QD} remains finite [$j_{QD} \to eN_S^{\min}/\tau_{QD}$, see (85)].

In (89), the quantity $N_S (1 - f_n)(1 - f_p)$ [where $f_{n,p}$ are given by (88)] increases with N_S. Hence, the temperature T_d increases with the surface density of QDs (Fig. 30). The point is that j_{QD} increases with N_S, whereas j_{OCL} decreases.

The temperature T_d increases with the bandgap difference between the materials of the OCL and QDs ΔE_{g1}. The reason is that the free carrier densities and hence j_{OCL} decrease with ΔE_{g1}.

The greater δ or β, the greater is $N_S^{\min}$ [see (32)] and hence the greater are $f_{n,p}$ [see (88)]. Hence, the temperature T_d decreases with increasing the RMS of relative QD size fluctuations δ or total losses β (Fig. 31).

With (86) and (89), the ratio of j_{OCL} to j_{QD} may be presented as

$$\frac{j_{OCL}(T)}{j_{QD}(T)} = \frac{j_{QD}(T_d) - \frac{eN_S^{\min}}{\tau_{QD}}}{j_{QD}(T) - \frac{eN_S^{\min}}{\tau_{QD}}} \left(\frac{T}{T_d}\right)^{3/2} \exp\left(\frac{\Delta E_{g1} - \varepsilon_n - \varepsilon_p}{T_d} - \frac{\Delta E_{g1} - \varepsilon_n - \varepsilon_p}{T}\right). \tag{90}$$

6.2. Characteristic temperature

With the above equations for the components of j_{th}, it is an easy matter to calculate the characteristic temperature of a QD laser — a very important parameter describing empirically the temperature dependence of j_{th} of semiconductor lasers

$$T_0 = \left(\frac{\partial \ln j_{th}}{\partial T}\right)^{-1}. \tag{91}$$

Although the T-dependence of j_{th} is never the exponential $j_0 \exp(T/T_0)$ [45] [as might appear from (91)], T_0 does characterize the temperature dependence of j_{th} adequately, provided it is the function of the temperature by itself: $T_0 = T_0(T)$.

The characteristic temperature of a QD laser may be presented as

$$\frac{1}{T_0} = \frac{j_{QD}}{j_{QD} + j_{OCL}} \frac{1}{T_0^{QD}} + \frac{j_{OCL}}{j_{QD} + j_{OCL}} \frac{1}{T_0^{OCL}} \tag{92}$$

where T_0^{QD} and T_0^{OCL} are defined similarly to T_0 for the functions $j_{QD}(T)$ and $j_{OCL}(T)$, respectively: $1/T_0^{QD} = \partial \ln j_{QD}/\partial T$ and $1/T_0^{OCL} = \partial \ln j_{OCL}/\partial T$. Hence, the reciprocal of T_0 is the sum of the reciprocals of T_0^{QD} and T_0^{OCL}, each weighted by the relative contribution of the respective component of j_{th}. With (90), the relative contributions are expressible as functions of T (Fig. 25).

With (85) and (88), T_0^{QD} becomes

$$\frac{1}{T_0^{\text{QD}}} = -\frac{1}{4}\frac{1}{f_n f_p}\frac{\partial \Delta^2}{\partial T}. \tag{93}$$

Since the absolute value of $\Delta\left(N_{\text{S}}/N_{\text{S}}^{\min}, T\right)$ decreases with T (Fig. 22), T_0^{QD} is,

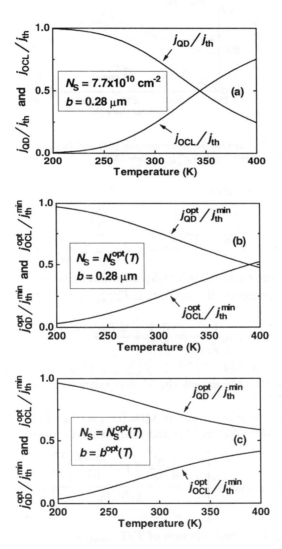

Fig. 25. Temperature dependences of the relative contributions of the components to the threshold current density for the structure optimized at $T = 300$ K (a) and to the minimum threshold current density for a fixed value of $b = 0.28\,\mu\text{m}$ (b) and for $b = b^{\text{opt}}(T)$ (c). Optimization with respect to b is of significance, provided optimization with respect to N_{S} is also carried out: if $b = b^{\text{opt}}(T)$, the contribution of the recombination in QDs is always greater than that in the OCL.

of course positive. With (88), (75), and (74) it is an easy matter to calculate T_0^{QD} (shown in Fig. 26). If the charge neutrality in QDs were the case ($\Delta = 0$), T_0^{QD} would be infinitely high.

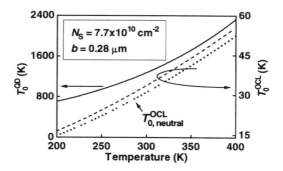

Fig. 26. T_0^{QD} and T_0^{OCL} against temperature [see (93) and (94)] for the structure optimized at $T = 300$ K. The dotted curve shows $T_{0,\text{neutral}}^{OCL}$ [see (95)].

With (86), T_0^{OCL} becomes

$$\frac{1}{T_0^{OCL}} = \frac{1}{T_{0,\text{neutral}}^{OCL}} - \frac{\frac{eN_S^{\min}}{T_{QD}}}{j_{QD} - \frac{eN_S^{\min}}{T_{QD}}} \frac{1}{T_0^{QD}} \quad (94)$$

where $T_{0,\text{neutral}}^{OCL}$ is the characteristic temperature calculated for the function $j_{OCL}(T)$ assuming the charge neutrality in QDs: $1/T_{0,\text{neutral}}^{OCL} = \partial \ln(Bn_1p_1)/\partial T$. With (87), $T_{0,\text{neutral}}^{OCL}$ becomes

$$\frac{1}{T_{0,\text{neutral}}^{OCL}} = \frac{3}{2}\frac{1}{T} + \frac{\Delta E_{g1} - \varepsilon_n - \varepsilon_p}{T^2}. \quad (95)$$

As may be seen from (94), $T_0^{OCL} > T_{0,\text{neutral}}^{OCL}$ (Fig. 26). Hence, the neutrality violation weakens to some extent the T–dependence of j_{OCL} [in comparison with that given by (87)].

Assuming the charge neutrality in QDs [i.e., setting $T_0^{QD} = \infty$ in (92) and (94) and $f_{n,p} = \frac{1}{2}(1 + N_S^{\min}/N_S)$ in (85) and (86)], we may present $T_{0,\text{neutral}}$ as

$$T_{0,\text{neutral}} = \left[1 + \frac{\frac{1}{4}\frac{N_S}{\tau_{QD}}\left(1 - \frac{N_S^{\min}}{N_S}\right)^2}{bBn_1p_1}\right] T_{0,\text{neutral}}^{OCL} \quad (96)$$

or as the following universal function of $T/T_{d,\text{neutral}}$:

$$\frac{T_{0,\text{neutral}}}{T_{d,\text{neutral}}} = \frac{\left(\frac{T}{T_{d,\text{neutral}}}\right)^2}{\frac{3}{2}\frac{T}{T_{d,\text{neutral}}} + \frac{\Delta E_{g1} - \varepsilon_n - \varepsilon_p}{T_{d,\text{neutral}}}}$$

$$\times \left\{ 1 + \left(\frac{T_{d,\text{neutral}}}{T}\right)^{3/2} \exp\left[\frac{\Delta E_{g1} - \varepsilon_n - \varepsilon_p}{T_{d,\text{neutral}}}\left(\frac{T_{d,\text{neutral}}}{T} - 1\right)\right] \right\}. \quad (97)$$

Here, $T_{d,\text{neutral}}$ is given by (89) wherein $f_{n,p} = \frac{1}{2}(1 + N_S^{\min}/N_S)$ are inserted.

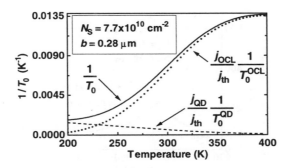

Fig. 27. $1/T_0$ and the first and the second terms in the right-hand side of (92) against T.

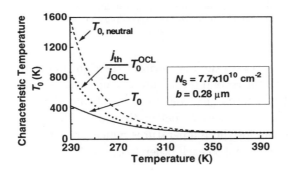

Fig. 28. Characteristic temperature T_0 against temperature. The dashed curve depicts T_0 calculated assuming the charge neutrality in QDs [see (96) and (97)].

The characteristic temperature of a QD laser depends strongly on T; T_0 falls off profoundly with increasing T (Figs. 27, 28, and 29). A drastic decrease in T_0 occurs in passing from temperature conditions wherein j_{th} is controlled by recombination in QDs to temperature conditions wherein it is controlled by recombination in the OCL (Fig. 25).

We emphasize that the tendency in T_0 to decrease drastically with T is in line with experimental results.[5]

It should be noted that T_0^{QD} is much greater than T_0^{OCL} (Fig. 26). Nevertheless, as may be seen from (92), $1/T_0$ is controlled not only by $1/T_0^{\text{QD}}$ and $1/T_0^{\text{OCL}}$, but by the relative contributions of the threshold current density components, $j_{\text{QD}}/(j_{\text{QD}} + j_{\text{OCL}})$ and $j_{\text{OCL}}/(j_{\text{QD}}+j_{\text{OCL}})$, as well. For this reason, *under temperature conditions wherein j_{th} is controlled by j_{QD} (for $T < T_d$), the contribution of the first term in the* right-hand side of (92) [i.e., of $(j_{\rm QD}/j_{\rm th})(1/T_0^{\rm QD})$] is every bit as important as that of the second term [i.e., of $(j_{\rm OCL}/j_{\rm th})(1/T_0^{\rm OCL})$] (Fig. 27). For such temperatures, eqs. (96) and (97) would give the $T_{0,\,{\rm neutral}}$ value (Fig. 28) that would be far in excess of the actual T_0 given by (92) and taking into account violation of the charge neutrality in QDs. Hence, eqs. (96) and (97) are completely inapplicable for $T < T_{\rm d}$.

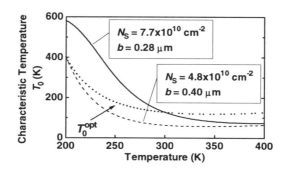

Fig. 29. Characteristic temperature T_0 against temperature for the structures optimized at $T = 300$ K and $T = 200$ K (the solid and dashed curves, respectively). The dotted curve plots $T_0^{\rm opt}$ [see (102)]. The solid, dashed and dotted curves in the figure correspond to the solid, dashed and dotted curves in Fig. 33. *For the temperatures exceeding that at which the structure is optimized, the dotted curve marks the upper bound for the set of curves for $T_0(T)$ of a QD laser.*

As well as giving rise to the first term in the right-hand side of (92), neutrality violation also affects the second term (through affecting $j_{\rm QD}$, $j_{\rm OCL}$ and $T_0^{\rm OCL}$). The reciprocal of the latter term, $(j_{\rm th}/j_{\rm OCL})\,T_0^{\rm OCL}$, calculated having regard to violation of the neutrality approximates T_0 better than does $T_{0,\,{\rm neutral}}$ (Fig. 28). However it is also inapplicable for the temperatures which are less than that at which the terms in (92) become equal to each other ($\partial j_{\rm OCL}/\partial T = \partial j_{\rm QD}/\partial T$).

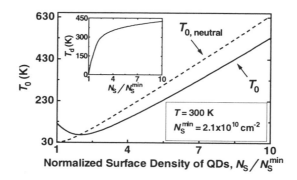

Fig. 30. Characteristic temperature T_0 against normalized surface density of QDs. The dashed curve depicts $T_{0,\,{\rm neutral}}$ [see (96) and (97)]. The nonmonotonic character of the initial portion of the dependence of T_0 on $N_{\rm S}$ is caused by violation of the charge neutrality in QDs. The inset shows the temperature $T_{\rm d}$ at which the components of $j_{\rm th}$ become equal to each other.

As may be seen from (96) and Fig. 30, $T_{0,\text{neutral}}$ increases monotonically with N_S. For N_S fairly greater than $N_S^{\min}$, the actual T_0 also increases with N_S (Fig. 30). The point is that the less temperature-sensitive component of j_{th}, i.e., j_{QD} increases with N_S, whereas the more sensitive component, i.e., j_{OCL} decreases.

The greater the RMS of relative QD size fluctuations δ or total losses β, the less is T_0 at given T and given other parameters (Fig. 31). The reason is the greater δ or β, the less perfect is the structure.

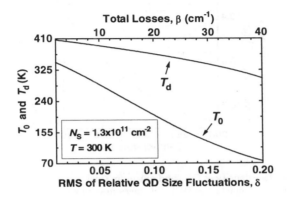

Fig. 31. Characteristic temperature T_0 and the temperature T_d against RMS of relative QD size fluctuations δ for $\beta = 10\,\text{cm}^{-1}$ (the bottom axis) and the total losses β for $\delta = 0.05$ (the top axis). N_S is taken so that it exceeds $N_S^{\min}$ throughout the entire range of δ displayed ($N_S^{\min} \approx 8.3 \times 10^{10}\,\text{cm}^{-2}$ for $\delta = 0.2$).

6.3. *Temperature dependence of j_{th} and T_0 for optimized structures*

The optimum parameters of the QD structure are temperature dependent (Fig. 32). What this means is for every operating temperature there is a specific optimized structure of its own (possessing the minimum threshold current density at this T). Hence, the laser optimized at a given T is not optimized at any other T.

The optimum thickness of the OCL decreases with T (Fig. 32). The decrease of b^{opt} tends to compensate to some extent the increase of the recombination rate in the OCL, Bnp, caused by the increase of the free carrier densities, n and p.

The optimum surface density of QDs increases with T (Fig. 32). This is due to the following. Because of increase of the thermoactivative leakage current (recombination current in the OCL) increase of N_S is required to ensure lasing.

For the structure optimized at a given T, the components of $j_{\text{th}}^{\min}$ associated with the radiative recombination in QDs and in the OCL are

$$j_{\text{QD}}^{\text{opt}} = \frac{eN_S^{\min}(b^{\text{opt}})}{\tau_{\text{QD}}} + \sqrt{\frac{eN_S^{\min}(b^{\text{opt}})}{\tau_{\text{QD}}} eb^{\text{opt}} Bn_1 p_1} \qquad (98)$$

158 L. V. Asryan & R. A. Suris

$$j_{\text{OCL}}^{\text{opt}} = eb^{\text{opt}} B n_1 p_1 + \sqrt{\frac{eN_S^{\min}(b^{\text{opt}})}{\tau_{\text{QD}}} eb^{\text{opt}} B n_1 p_1} \qquad (99)$$

[see Fig. 25(c)]. Equations (98) and (99) will also remain valid if optimization with respect to only N_S is carried out [see Fig. 25(b)]. In that case, the given value of b enters into them instead of $b^{\text{opt}}(T)$.

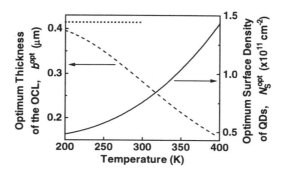

Fig. 32. Optimum surface density of QDs and optimum thickness of the OCL against temperature. The horizontal dotted line depicts the OCL thickness maximizing the optical confinement factor.

Fig. 33 plots $j_{\text{th}}^{\min} = j_{\text{QD}}^{\text{opt}} + j_{\text{OCL}}^{\text{opt}}$ (the dotted curve) obtained by optimizing with respect to both N_S and b. Each point on the dotted curve corresponds to the specific structure with the optimum values of N_S and b at the given T. At given δ and β, the dotted curve marks the lower bound for the set of curves for $j_{\text{th}}(T)$ of QD laser. For T beyond the temperature at which the structure is optimized (the point of tangency of the dotted and solid curves or of the dotted and dashed curves in the figure), j_{th} increases with T more steeply than does $j_{\text{th}}^{\min}$.

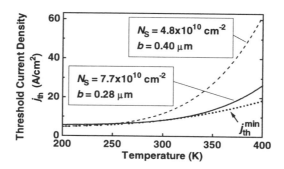

Fig. 33. Threshold current density against temperature for the structures optimized at $T = 300$ and 200 K (the solid and dashed curves, respectively). The dotted curve plots $j_{\text{th}}^{\min}$ obtained by optimizing with respect to both N_S and b. The solid, dashed and dotted curves in the figure correspond to the solid, dashed and dotted curves in Fig. 29.

With (98) and (99), wherein the given value of b enters instead of $b^{\text{opt}}(T)$, the characteristic temperature calculated for the function $j_{\text{th}}^{\min}(T)$ obtained by optimizing with respect to only N_S is

$$T_0^{\text{opt}} = \left[1 + \left(\frac{\frac{N_S^{\min}}{\tau_{\text{QD}}}}{bBn_1p_1}\right)^{1/2}\right] T_{0,\text{neutral}}^{\text{OCL}} \qquad (100)$$

[contrast (100) with (96)]. An equation similar to (89) can be obtained for the temperature T_d^{opt} at which the components of $j_{\text{th}}^{\min}$ (obtained by optimizing with respect to only N_S) become equal to each other, $j_{\text{OCL}}^{\text{opt}}(T_d^{\text{opt}}) = j_{\text{QD}}^{\text{opt}}(T_d^{\text{opt}})$ [Fig. 25(b)]. The only difference between this equation and (89) is the following: in the equation for T_d^{opt}, the denominator of the argument of the logarithmic function is $N_S^{\min}$ instead of $N_S(1 - f_n)(1 - f_p)$ in (89). With T_d^{opt}, T_0^{opt} may be presented as the following universal function of T/T_d^{opt}:

$$\frac{T_0^{\text{opt}}}{T_d^{\text{opt}}} = \frac{\left(\frac{T}{T_d^{\text{opt}}}\right)^2}{\frac{3}{2}\frac{T}{T_d^{\text{opt}}} + \frac{\Delta E_{g1} - \varepsilon_n - \varepsilon_p}{T_d^{\text{opt}}}}$$

$$\times \left\{1 + \left(\frac{T_d^{\text{opt}}}{T}\right)^{3/4} \exp\left[\frac{\Delta E_{g1} - \varepsilon_n - \varepsilon_p}{2T_d^{\text{opt}}}\left(\frac{T_d^{\text{opt}}}{T} - 1\right)\right]\right\}. \qquad (101)$$

Contrast (101) with (97); note that a half of the energy of excitation of the electron-hole pair from a QD, $\frac{1}{2}(\Delta E_{g1} - \varepsilon_n - \varepsilon_p)$, enters into the argument of the exponential in (101). In Fig. 25(b), $T_d^{\text{opt}} = 390$ K.

When optimization with respect to both N_S and b is carried out, $j_{\text{QD}}^{\text{opt}}(T) > j_{\text{OCL}}^{\text{opt}}(T)$ at any T. With $T \to \infty$, $eb^{\text{opt}}(T)B(T)n_1(T)p_1(T) \to eN_S^{\min}[b^{\text{opt}}(T)]/\tau_{\text{QD}}$ and the components of $j_{\text{th}}^{\min}$ will tend to each other: $j_{\text{OCL}}^{\text{opt}} \to j_{\text{QD}}^{\text{opt}}$ [see (98) and (99) and Fig. 25(c)]. Hence, $T_d^{\text{opt}} = \infty$ for this case.

The characteristic temperature calculated for the function $j_{\text{th}}^{\min}(T)$ obtained by optimizing with respect to both N_S and b is

$$\frac{1}{T_0^{\text{opt}}} = \left\{\frac{1}{1 + \left[\frac{a}{b^{\text{opt}}\Gamma(b^{\text{opt}})^s}\right]^{1/2}}\left(1 - \frac{d\ln b^{\text{opt}}}{d\ln s}\right)\right.$$

$$\left. + \frac{\left[\frac{a}{b^{\text{opt}}\Gamma(b^{\text{opt}})^s}\right]^{1/2}}{1 + \left[\frac{a}{b^{\text{opt}}\Gamma(b^{\text{opt}})^s}\right]^{1/2}}\frac{d\ln \Gamma(b^{\text{opt}})}{d\ln s}\right\}\frac{1}{T_{0,\text{neutral}}^{\text{OCL}}}. \qquad (102)$$

Expression in the brackets is the universal function of the dimensionless parameter s [see (45)] and can be obtained from the s-dependences of b^{opt} and $\Gamma(b^{\text{opt}})$ (Fig. 7). Both $d\ln b^{\text{opt}}/d\ln s$ and $d\ln\Gamma(b^{\text{opt}})/d\ln s$ are positive and less than $\frac{1}{2}$; they tend to 0 and $\frac{1}{2}$ as $s \to \infty$ ($T \to 0$) and $s \to 0$ ($T \to \infty$), respectively. With $T \to \infty$ ($s \to 0$), $sa/[b^{\text{opt}}\Gamma(b^{\text{opt}})] \to 1$ [eq. (57) in Ref.[25]] and $T_0^{\text{opt}} \to 2\,T_{0,\text{neutral}}^{\text{OCL}}$. The dotted curve in Fig. 29 plots T_0 calculated from (102). Each point on this curve corresponds to the specific structure with the optimum values of N_S and b at the given T. For the temperatures exceeding that at which the structure is optimized ($T = 300$ K and $T = 200$ K for the solid and dashed curves respectively), T_0 of the real structure can not be higher than T_0^{opt} given by (102). The same is apparent from Fig. 33, whence it follows that for T beyond the temperature of the point of tangency of the dotted and solid curves or of the dotted and dashed curves (correspondingly 300 K and 200 K), j_{th} increases with T more steeply than does $j_{\text{th}}^{\min}(T)$. Therefore, T_0 should be less than T_0^{opt} calculated for $j_{\text{th}}^{\min}(T)$.

7. Longitudinal Spatial Hole Burning

The problem of multimode generation is of first importance for laser applications.[41] A study of the physical processes controlling the multimode generation threshold is necessary to find ways of suppressing the additional modes and to offer a proper design of single-mode operating lasers.

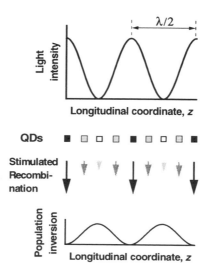

Fig. 34. Light intensity and population inversion in QDs along the longitudinal direction in the waveguide.

As in conventional semiconductor lasers,[46,47] spatial hole burning (SHB) in QD lasers is due to nonuniformity of stimulated recombination of carriers along the longitudinal direction in the waveguide. The point is that, at and above the lasing threshold, the emitted light is a standing wave (and several thousands of halfwavelengths are accommodated in the cavity length). At the same time, several QDs are arranged within a halfwavelength in the medium $\lambda_0/(2\sqrt{\epsilon})$ (period of the light intensity). For the typical surface density of QDs ($N_S = 3 \times 10^{10}$—10^{11} cm^{-2}), the mean separation between QD centers, $d = 1/\sqrt{N_S}$, is only several hundreds of ångstroms, while $\lambda_0/(2\sqrt{\epsilon})$ is more than 2000 Å for the most actual for telecommunication needs wavelengths ($\lambda_0 = 1.3$ and $1.55\,\mu$m). Hence, the stimulated recombination will be most intensive in those QDs located at the antinodes of the light intensity, while it will be lest intensive in the QDs located at the nodes (Fig. 34). As a result, depletion of the QDs located near the antinodes and overfilling of those located near the nodes may take place. This leads to the lasing generation of the higher longitudinal modes.

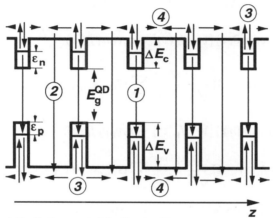

1. Radiative recombination in QDs
2. Band-to-band radiative recombination in the OCL
3. Capture into QDs and thermally excited escape from QDs
4. Diffusion along the longitudinal direction

Fig. 35. Schematic of a QD laser structure along the longitudinal direction and the main processes.

In QW or bulk lasers, diffusion in the active region will tend to smooth out the nonuniform carrier distribution and population inversion along the longitudinal direction, thus suppressing totally or partly the SHB effect.[46]

A drastically different situation occurs in QD lasers. Here, diffusion will play a similar yet minor role. The point is that the carriers contributing to the stimulated emission are those totally confined in QDs and hence the smoothing-out of the spatially nonuniform population inversion may be strongly suppressed. There are also free carriers in the optical confinement layer (OCL), which contribute to the spontaneous emission. The free-carrier densities and confined carrier level occupancies are coupled to each other through the processes of carrier capture into QDs and thermally excited escape from QDs, and free-carrier diffusion (Fig. 35). Due to this coupling, diffusion along the longitudinal direction should equalize to some extent the level occupancies in different QDs. Hence, the slowest of the capture–escape and diffusion processes controls the spatial distribution of free and confined carriers. In this section, thermal excitations from QDs, rather than diffusion, are shown to limit the smoothing-out of the spatially nonuniform population inversion; nonvanishing values of the characteristic times of thermally excited escapes govern the multimode generation threshold. A very similar situation had been considered in Refs.[48,49] for band-to-impurity transitions lasing.

7.1. Rate equations

The steady-state rate equations for the carriers confined in QDs, free carriers and photons, $\partial (f_n, f_p, n, p, N_{m_l})/\partial t = 0$, are as follows:

$$\sigma_n v_n n(0,z) [1 - f_n(z)] - \sigma_n v_n n_1 f_n(z) - \frac{f_n(z) f_p(z)}{\tau_{QD}}$$

$$- [f_n(z) + f_p(z) - 1] \frac{c}{\sqrt{\epsilon_g}} \frac{g^{max}}{N_S S} \sum_{m_l} G_{m_l} N_{m_l} (1 + \cos 2k_{m_l} z) = 0 \qquad (103)$$

$$\sigma_p v_p p(0,z) [1 - f_p(z)] - \sigma_p v_p p_1 f_p(z) - \frac{f_n(z) f_p(z)}{\tau_{QD}}$$

$$- [f_n(z) + f_p(z) - 1] \frac{c}{\sqrt{\epsilon_g}} \frac{g^{max}}{N_S S} \sum_{m_l} G_{m_l} N_{m_l} (1 + \cos 2k_{m_l} z) = 0 \qquad (104)$$

$$\sigma_n v_n n_1 N_S f_n(z) \delta(x) - \sigma_n v_n n(x,z) N_S [1 - f_n(z)] \delta(x)$$

$$- B n(x,z) p(x,z) + D_n \left[\frac{\partial^2 n(x,z)}{\partial x^2} + \frac{\partial^2 n(x,z)}{\partial z^2} \right] = 0 \qquad (105)$$

$$\sigma_p v_p p_1 N_S f_p(z) \delta(x) - \sigma_p v_p p(x,z) N_S [1 - f_p(z)] \delta(x)$$

$$- B n(x,z) p(x,z) + D_p \left[\frac{\partial^2 p(x,z)}{\partial x^2} + \frac{\partial^2 p(x,z)}{\partial z^2} \right] = 0 \qquad (106)$$

$$N_{m_l} \frac{c}{\sqrt{\epsilon_g}} \int_0^L g^{max} G_{m_l} [f_n(z) + f_p(z) - 1] (1 + \cos 2k_{m_l} z) \frac{dz}{L} - N_{m_l} \frac{c}{\sqrt{\epsilon_g}} \beta_{m_l} = 0$$

$$(107)$$

where $S = WL$ is the surface area of the QD layer, W is the QD layer width (the lateral size), L is the QD layer length (the cavity length), $D_{n,p}$ are the diffusion constants, $\sqrt{\epsilon_g}$ is the group index of the dispersive OCL material, and β_{m_l} is the mirror loss for the m_l-th longitudinal mode.

In (103)–(107), $f_{n,p}(z)$ are the mean (averaged over the lateral direction) electron and hole level occupancies in QDs; $n(x,z)$ and $p(x,z)$ are the free carrier densities in the OCL dependent on both the transverse coordinate x (direction of the current injection) and longitudinal coordinate z (direction of the emitted light yield). It is SHB that causes the z-dependence of $f_{n,p}$, n and p.

The following assumptions and simplifications are made here.

(i) Since the QD size is much less than the OCL thickness, the active layer with QDs can be considered as a δ-layer (the plain $x = 0$), hence δ-function enters into (105) and (106).

(ii) We consider an ideal situation when there is only one electron and one hole energy level in a QD (which are calculated using a simplified model in Ref.[25]). In such a situation, the expected advantages of QD lasers over the conventional QW ones will be pronounced most strongly. In actual laser structures, there are also excited states in a QD. We examined the effect of excited-state transitions on the threshold characteristics of a laser in Refs.[50,51].

(iii) For properly designed QD lasers (when parameters are well away from the critical values – see Section 4.2), the internal losses, being mainly due to free carriers, should be strongly suppressed. This is because the free-carrier densities in such structures are much less (by one or even two orders of magnitude – see Fig. 23) than those in QW lasers. This fact stems from the δ-function like density of states for carriers in QDs and the high material gain of a laser. For this reason, the internal losses are neglected compared to the mirror ones.

(iv) Calculation of the capture cross sections $\sigma_{n,p}$ is beyond the scope of this work. Here, to estimate the multimode generation threshold, $\sigma_{n,p}$ are plausibly taken to be 10^{-13} cm^2 (which is much less than the geometrical cross section of a QD). We emphasize that the character of the dependence of the multimode generation threshold on structure parameters and temperature obtained here will remain unchanged with $\sigma_{n,p}$.

(v) A set of eqs. (103)–(107) is general. Nevertheless, since our prime interest here is in finding ways of suppressing the additional modes, the condition for the lasing oscillation of the next (closest) to the main mode is examined comprehensively (actually these may be two modes symmetrical about the main one if the gain spectrum is symmetrical about its maximum). Hence we consider injection currents which are above the threshold current for the main mode and below that for the next one (relatively small power outputs).

The first term in the left-hand side of (103) and (104) is the flux of the electron or hole capture into QDs, respectively. The second term is the flux of thermally excited escapes from QDs; this flux can be rewritten as $N_S f_{n,p}(z)/\tau_{n,p}^{esc}$ where the characteristic times of thermally excited escapes are given by (10).

The third term in (103) and (104) is the flux of the spontaneous radiative recombination in QDs. The last term is the flux of the stimulated recombination in QDs. The summation is over all modes contributing to the stimulated emission. The dimensionless quantity G_{m_l} is the line shape factor of the gain for the m_l-th longitudinal mode, and N_{m_l} is the number of intracavity photons in this mode. The z-dependence of the intensity of the m_l-th mode is accounted for by the factor $1 + \cos 2k_{m_l} z$ (which reflects the standing-wave nature of the mode), where $k_{m_l} = 2\pi\sqrt{\epsilon}/\lambda_{m_l}$. For the Fabri-Perot cavity of length L, the supported wavelengths are $\lambda_{m_l} = (2L/m_l)\sqrt{\epsilon}$, where m_l is an integer.

As in Ref.[46], we separated out the z-dependent population inversion in QDs, $f_n(z) + f_p(z) - 1$, from the equation for the gain. The modal gain spectrum is

$$g(E) = g^{\max}(f_n + f_p - 1)G(E) \tag{108}$$

with the line shape factor spectrum $G(E) = w[(\bar{E}_0 - E)/(q_n \bar{\epsilon}_n + q_p \bar{\epsilon}_p)]/w(0)$, where the function w is the probability density of relative QD size fluctuations, $\bar{E}_0$ is the energy of the gain spectrum maximum. The equation for the maximum gain $g^{\max}$ is given by (13). The line shape factor of the gain for the m_l-th mode is $G_{m_l} = G(2\pi\hbar c/\lambda_{m_l})$.

The first and the second terms in the left-hand sides of (105) and (106) are the rates of the thermally excited escapes from QDs and capture into QDs, respectively. The third term is the spontaneous radiative recombination rate in the OCL. The last term is the free carrier diffusion rate.

With (103) and (104), the free carrier densities in the OCL nearby the QD layer are readily expressed in terms of $f_{n,p}(z)$ to yield

$$n(0,z) = n_1 \frac{f_n(z)}{1 - f_n(z)} + \frac{1}{\sigma_n v_n \tau_{QD}} \frac{f_n(z) f_p(z)}{1 - f_n(z)}$$

$$+ \frac{1}{\sigma_n v_n} \frac{f_n(z) + f_p(z) - 1}{1 - f_n(z)} \frac{c}{\sqrt{\epsilon_g}} \frac{g^{\max}}{N_S S} \sum_{m_l} G_{m_l} N_{m_l} (1 + \cos 2k_{m_l} z) \tag{109}$$

and a similar equation for $p(0,z)$. Conversely, $f_{n,p}(z)$ are expressible analytically in terms of $n(0,z)$ and $p(0,z)$ [a set of eqs. (103) and (104) can be rearranged into the quadratic equations in $f_{n,p}(z)$].

The injection current density flowing across the junction (along the x-direction) is

$$j = eB \int_{-b/2}^{b/2} \langle np \rangle dx + \frac{eN_S}{\tau_{QD}} \langle f_n f_p \rangle$$

$$+ e\frac{1}{S} \langle (f_n + f_p - 1) \frac{c}{\sqrt{\epsilon_g}} g^{\max} \sum_{m_l} G_{m_l} N_{m_l} (1 + \cos 2k_{m_l} z) \rangle \tag{110}$$

where b is the OCL thickness, and $\langle ... \rangle$ means averaging in the z-direction.

The last terms in (109) and (110) are associated with the stimulated recombination. It should be emphasized that, above the lasing threshold, – due to SHB – the first two terms also differ from their threshold values.

The first term in the right-hand side of (107) describes the stimulated emission into the m_l-th mode, and the second term gives the damping of that mode. Integrating means averaging (over the z-direction) of the gain for the m_l-th mode weighted by the factor $1 + \cos 2k_{m_l} z$ presenting the spatial modulation of the light intensity of the mode.

Eq. (107) may be rewritten as

$$g^{\max} G_{m_l} \int_0^L [f_\mathrm{n}(z) + f_\mathrm{p}(z) - 1](1 + \cos 2k_{m_l} z) \frac{\mathrm{d}z}{L} = \beta_{m_l} \qquad (111)$$

which is simply the extension of the threshold condition for the above threshold regime.

It is evident from (111) that the average (over the z-direction) level occupancies and population inversion in QDs are built up compared to the threshold values due to SHB (the factor $1 + \cos 2k_{m_l} z$); and the same is true for the average free carrier densities.

Instead of (105) and (106), the equations

$$D_\mathrm{n}\left(\frac{\partial^2 n}{\partial x^2} + \frac{\partial^2 n}{\partial z^2}\right) = Bnp \qquad D_\mathrm{p}\left(\frac{\partial^2 p}{\partial x^2} + \frac{\partial^2 p}{\partial z^2}\right) = Bnp \qquad (112)$$

can be equivalently used separately for the left- and right-hand (with respect to the QD layer) sides of the OCL (the regions $-b/2 \leq x < 0$ and $0 < x \leq b/2$, respectively) provided they are supplemented with the following boundary conditions obtained by integrating (105) and (106) over an infinitesimal interval around $x = 0$ (actually, over the QD layer $-a/2 \leq x \leq a/2$) and using (103) and (104):

$$D_\mathrm{n}\left(\left.\frac{\partial n}{\partial x}\right|_{x=+0} - \left.\frac{\partial n}{\partial x}\right|_{x=-0}\right) = N_\mathrm{S}\frac{f_\mathrm{n}(z)f_\mathrm{p}(z)}{\tau_\mathrm{QD}}$$

$$+\frac{1}{S}[f_\mathrm{n}(z) + f_\mathrm{p}(z) - 1]\frac{c}{\sqrt{\epsilon_\mathrm{g}}} g^{\max} \sum_{m_l} G_{m_l} N_{m_l} (1 + \cos 2k_{m_l} z) \qquad (113)$$

$$D_\mathrm{p}\left(\left.\frac{\partial p}{\partial x}\right|_{x=+0} - \left.\frac{\partial p}{\partial x}\right|_{x=-0}\right) = N_\mathrm{S}\frac{f_\mathrm{n}(z)f_\mathrm{p}(z)}{\tau_\mathrm{QD}}$$

$$+\frac{1}{S}[f_\mathrm{n}(z) + f_\mathrm{p}(z) - 1]\frac{c}{\sqrt{\epsilon_\mathrm{g}}} g^{\max} \sum_{m_l} G_{m_l} N_{m_l} (1 + \cos 2k_{m_l} z). \qquad (114)$$

Eq. (113) [and (114)] has an evident meaning: the difference in the electron [and hole] current density at the right- and left-hand side heteroboundaries of the OCL and QD-layer goes into the spontaneous and stimulated recombination in QDs.

7.2. Multimode generation threshold

At and slightly above the lasing threshold, only the main (closest to the gain spectrum maximum) mode oscillates. The number of the lasing modes increases with the injection current [see (110)].

We define the multimode generation threshold δj as the excess of the injection current density over the threshold current density for the main mode j_th required for oscillating the next longitudinal mode:

$$\delta j = j_2 - j_\mathrm{th} \qquad (115)$$

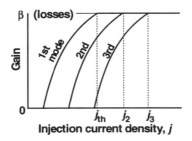

Fig. 36. Gain (below the lasing threshold) or averaged (over z-coordinate) product of the gain and the light-intensity spatial variation function (above the lasing threshold) for the first three longitudinal modes against injection current density. $j_{\rm th}$, j_2 and j_3 are the threshold current densities for the 1st, 2nd and 3rd modes.

where j_2 is the threshold current density for the next mode (Fig. 36).

An examination of the problem yields [31]

$$\delta j = 2(f_{n0} + f_{p0} - 1)\,\delta G$$

$$\times \frac{1 + \left[f_{p0}\dfrac{n_1}{n_1+n_0(0)}\left(\dfrac{1}{j_n^{\rm esc}}+\dfrac{1}{j_n^{\rm D}}\right) + f_{n0}\dfrac{p_1}{p_1+p_0(0)}\left(\dfrac{1}{j_p^{\rm esc}}+\dfrac{1}{j_p^{\rm D}}\right)\right]j_{\rm QD}^{\rm sp}}{\dfrac{n_1}{n_1+n_0(0)}\left(\dfrac{1}{j_n^{\rm esc}}+\dfrac{1}{j_n^{\rm D}}\right) + \dfrac{p_1}{p_1+p_0(0)}\left(\dfrac{1}{j_p^{\rm esc}}+\dfrac{1}{j_p^{\rm D}}\right)} \quad (116)$$

where $j_{n,p}^{\rm esc}$, $j_{n,p}^{\rm D}$, and $j_{\rm QD}^{\rm sp}$ are the current densities associated with thermally excited escapes, diffusion, and spontaneous radiative recombination, respectively:

$$j_{n,p}^{\rm esc} = \frac{eN_{\rm S}}{\tau_{n,p}^{\rm esc}} \qquad j_{\rm QD}^{\rm sp} = \frac{eN_{\rm S}}{\tau_{\rm QD}} \quad (117)$$

$$j_n^{\rm D} = 4ek_{m_1}D_n\frac{n_1}{1-f_{n0}}\tanh(k_{m_1}b) \qquad j_p^{\rm D} = 4ek_{m_1}D_p\frac{p_1}{1-f_{p0}}\tanh(k_{m_1}b). \quad (118)$$

In (116), $n_0(0)$ and $p_0(0)$ are the lasing threshold values of the free carrier densities in the OCL nearby the QD layer. They are given by the sums of the first two terms in (109) and in the equation for $p(0,z)$, wherein the threshold values, $f_{n0,p0}$, enter instead of $f_{n,p}(z)$.

The difference in the line shape factor of the main and the next modes (Fig. 37) is $\delta G = (1/2)\left|\partial^2 G/\partial E^2\right|(\delta E)^2$ where the derivative is taken at $E = E_0$, E_0 is photon energy of the main mode, $\delta E = \hbar(c/\sqrt{\epsilon})(\pi/L)$ is the separation between the photon energies of the neighbouring Fabri-Perot modes ($\Delta m_l = \pm 1$). For a Gaussian distribution of the relative QD size fluctuations,

$$\delta G = \frac{1}{2}\left(\frac{\hbar\dfrac{c}{\sqrt{\epsilon}}\dfrac{\pi}{L}}{(\Delta\varepsilon)_{\rm inhom}}\right)^2. \quad (119)$$

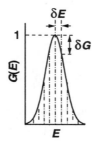

Fig. 37. Spectrum of the line shape factor of the gain.

For a Lorentzian distribution, the factor 1/2 in (119) should be dropped.

The ratio of the hole escape flux to the diffusion one is controlled by the cross section of the hole capture into a QD and by the surface density of QDs. (Since D_n is much greater than D_p, the free-electron diffusion is not the limiting factor; for GaInAsP, $D_n = 40\,\text{cm}^2/\text{s}$ and $D_p = 1.6\,\text{cm}^2/\text{s}$.[52]) There is only a weak temperature dependence of this ratio due to such dependence of the capture cross section, thermal velocity, and diffusion constant. The ratio is typically much less than unity:

$$\frac{j_p^{\text{esc}}}{j_p^D} = \frac{(1-f_{p0})N_S\sigma_p v_p}{4k_{m_1}D_p\tanh(k_{m_1}b)} \ll 1. \quad (120)$$

What this means is the process of thermally excited escapes from QDs is the slowest and hence the limiting one.

With (120), eq. (116) becomes

$$\delta j = 2(f_{n0}+f_{p0}-1)\,\delta G\,\frac{1+\left[f_{p0}\dfrac{n_1}{n_1+n_0(0)}\dfrac{1}{j_n^{\text{esc}}}+f_{n0}\dfrac{p_1}{p_1+p_0(0)}\dfrac{1}{j_p^{\text{esc}}}\right]j_{\text{QD}}^{\text{sp}}}{\dfrac{n_1}{n_1+n_0(0)}\dfrac{1}{j_n^{\text{esc}}}+\dfrac{p_1}{p_1+p_0(0)}\dfrac{1}{j_p^{\text{esc}}}}. \quad (121)$$

7.2.1. Relatively high temperatures

At relatively high temperatures (when equilibrium filling of QDs is the case below and at the lasing threshold), $\tau_{n,p}^{\text{esc}} \ll \tau_{\text{QD}}$ (and hence, equivalently, $j_{n,p}^{\text{esc}} \gg j_{\text{QD}}^{\text{sp}}$), $n_0(0) = n_1 f_{n0}/(1-f_{n0})$ and $p_0(0) = p_1 f_{p0}/(1-f_{p0})$ [see (16)].

The multimode generation threshold is controlled by the reciprocals of the characteristic times of thermally excited escapes from QDs:

$$\delta j = 2(f_{n0}+f_{p0}-1)\,\delta G\,\frac{eN_S}{\tau_n^{\text{esc}}(1-f_{n0})+\tau_p^{\text{esc}}(1-f_{p0})}. \quad (122)$$

The relative multimode generation threshold is defined as the ratio of δj to j_{th}.

With (122) and (20),

$$\frac{\delta j}{j_{\text{th}}} = 2(f_{n0} + f_{p0} - 1)\,\delta G\, \frac{\dfrac{\tau_{\text{QD}}}{\tau_n^{\text{esc}}(1-f_{n0}) + \tau_p^{\text{esc}}(1-f_{p0})}}{f_{n0}f_{p0} + \dfrac{\tau_{\text{QD}}}{N_{\text{S}}}bBn_1p_1\dfrac{f_{n0}f_{p0}}{(1-f_{n0})(1-f_{p0})}}. \tag{123}$$

Notice that typically $\delta G \ll 1$. Simultaneously, $\tau_{n,p}^{\text{esc}} \ll \tau_{\text{QD}}$. For the specific structure considered below, $\tau_{\text{QD}} = 0.7$ ns, and, at room temperature, $\tau_n^{\text{esc}} = 7$ ps, and $\tau_p^{\text{esc}} = 60$ ps.

As is seen from (10), (122) and (123), δj and $\delta j/j_{\text{th}}$ increase with $\sigma_{n,p}$. Clearly the greater $\sigma_{n,p}$, the more intensive the thermal excitations, and the less manifested the SHB effect.

Eq. (116) can also be used to roughly estimate the multimode generation threshold in QW lasers and then to compare with QD ones. To do this, the escape terms should be simply dropped in (116). Assuming the same material system, and that the population inversion and the difference in the line shape factor of the neighbouring modes in a QW and QD lasers are the same, we arrive at (120) for a lower limit to the ratio of δj in a QD laser to that in a QW one. This fact has an evident meaning: the hole diffusion controls the smoothing-out of the spatially nonuniform population inversion in a QW laser, whereas the hole escape controls in a QD one. For the GaInAsP/InP structure considered below, j_p^{esc}/j_p^D is 0.023. Taking into account the narrower gain spectrum [the larger δG – see (119)] of a properly designed QD laser as compared to a QW laser somewhat enhances the above ratio of the multimode generation thresholds.

With (37), δj at relatively high temperatures [eq. (122)] can be explicitly expressed as functions of N_{S}, δ and L to give:

$$\delta j(N_{\text{S}}) = 4\,\delta G\,\frac{eN_{\text{S}}^{\min}}{\tau_n^{\text{esc}} + \tau_p^{\text{esc}}}\,\frac{1}{1 - \dfrac{N_{\text{S}}^{\min}}{N_{\text{S}}}} \qquad (N_{\text{S}} > N_{\text{S}}^{\min}) \tag{124}$$

$$\delta j(\delta) = 2\left[\frac{\delta E}{(q_n\varepsilon_n + q_p\varepsilon_p)\delta}\right]^2 \frac{\delta}{\delta^{\max}}\frac{1}{1 - \dfrac{\delta}{\delta^{\max}}}\frac{eN_{\text{S}}}{\tau_n^{\text{esc}} + \tau_p^{\text{esc}}} \tag{125}$$

$$\delta j(L) = 2\left[\frac{\hbar\dfrac{c}{\sqrt{\epsilon}}\dfrac{\pi}{L}}{(\Delta\varepsilon)_{\text{inhom}}}\right]^2 \frac{L^{\min}}{L}\frac{1}{1 - \dfrac{L^{\min}}{L}}\frac{eN_{\text{S}}}{\tau_n^{\text{esc}} + \tau_p^{\text{esc}}} \qquad (L > L^{\min}). \tag{126}$$

In (125) and (126), a Gaussian distribution of the relative QD size fluctuations is assumed.

When one of structure parameters is close to its critical tolerable value, $f_{n0,p0}$ tend to unity [see (35)]. This demands infinitely high free-carrier densities in the OCL. As a result, both j_{th} and j_2, as well as δj, increase infinitely [see (122), (124)–(126) and Fig. 5]. As this takes place, $\delta j/j_{\text{th}}$ approaches zero [the insets in Fig. 5], which means the lasing generation of infinitely large number of longitudinal modes. This is because $\delta j \propto (1 - f_{n0})^{-1}$ [see (122)], whereas $j_{th} \propto (1 - f_{n0})^{-2}$ when $f_{n0,p0} \to 1$ [see equations (27) and (20) for j_{th}]. For the advantages of a QD laser to be attained in practice, structure parameters should be well away from the critical values.

As $\delta \to 0$, both δj and $\delta j/j_{th} \propto \delta^{-1}$ and increase infinitely [see (125) and Fig. 5(b)]; hence also j_2 increases infinitely [Fig. 5(b)].

As $L \to \infty$, both δj and $\delta j/j_{th} \propto L^{-3}$ and decay [see (126) and (Fig. 5(c)]; the threshold current density for the second mode, j_2, decreases and tends to that for the main mode, j_{th}, and hence also approaches the transparency current density [Fig. 5(c)].

As can be shown from (116), $\delta j \to (eN_S^{min}/\tau_{QD})\delta G = $ const with $N_S \to \infty$. [Notice, that inequality (120) does not hold and hence eqs. (121)–(127), (124), (125) and (126), derived using this inequality, are inapplicable for $N_S \to \infty$]. With this fact and the fact that, for such N_S, j_{th} is close to the transparency current density and hence increases linearly with N_S, we get the conclusion that $\delta j/j_{th}$ should decay as $N_S \to \infty$ [Fig. 5(a)]. Since $\delta j/j_{th} \to 0$ also as $N_S \to N_S^{min}$, there should be a certain value of N_S providing the maximum to $\delta j/j_{th}$ [Fig. 5(a)].

Since $\delta j/j_{th} \to 0$ both with $L \to L^{min}$ and $L \to \infty$, there is also a certain value of L at which $\delta j/j_{th}$ is a maximum [Fig. 5(c)].

Such a behaviour of the dependence of $\delta j/j_{th}$ on N_S and L offers ways to optimize the QD laser, aimed at maximizing the relative multimode generation threshold.

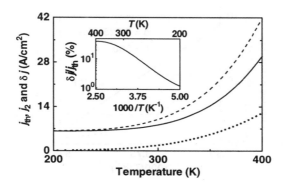

Fig. 38. Threshold current densities for the main (the solid curve) and the next (the dashed curve) longitudinal modes and the multimode generation threshold (the dotted curve) against temperature. The inset shows the relative multimode generation threshold.

The temperature dependence of the multimode generation threshold is caused by such dependence of the characteristic times of thermally excited escapes, $\tau_{n,p}^{esc}$ [see (10)]. The main reason is the exponential dependence of the quantities n_1 and p_1 on T; there is also a weak temperature dependence of the capture cross sections, $\sigma_{n,p}$, and thermal velocities, $v_{n,p}$. Concurrent with the increase of j_{th} and j_2, an increase of δj and $\delta j/j_{\text{th}}$ occurs with a rise in T (Fig. 38). This is because the

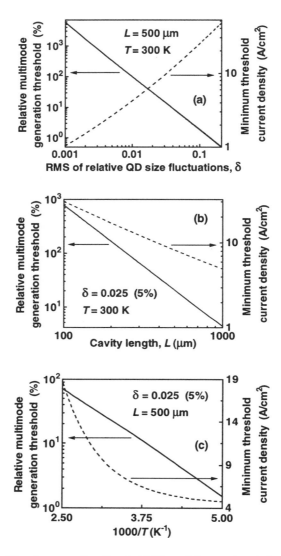

Fig. 39. Relative multimode generation threshold (the solid curves) and the minimum threshold current density for the main mode (the dashed curves) against RMS of relative QD size fluctuations (a), cavity length (b), and temperature (c). Each point on the curves corresponds to the specific structure optimized at the given δ (a), L (b) and T (c), respectively.

thermally excited escapes from QDs, and hence a smoothing-out of the carrier space distribution, become much more effective at high T. (In bulk semiconductor lasers too, δj increases with T due to the fact that the diffusion constant is higher at higher T.[46]) Thus, provided SHB is the only (or the main) factor allowing many modes to oscillate simultaneously in a QD laser, the number of the lasing modes decreases and hence the curve for the output power against injection current becomes more linear with T. This may be one of the reasons for an increase in the slope efficiency of a QD laser with T observed in Ref.[53].

The Coulomb interaction between free electrons and holes makes the diffusion process ambipolar-like. Since the ambipolar diffusion is faster than the hole diffusion, all the results obtained using inequality (120) and conclusions drawn will remain valid and unaltered if the Coulomb interaction is taken properly into account.

As our example, we used the same GaInAsP/InP heterojunction considered in Section 4.3.4. A device with OCL thickness of $b = 0.28\,\mu\text{m}$ and an as-cleaved facet at both ends is considered. A Gaussian distribution of the relative QD size fluctuations is assumed. The mean size of cubic QDs is taken to be $a = 150\,\text{Å}$. The surface density of QDs, RMS of relative QD size fluctuations, cavity length, and temperature are taken to be $N_S = 6.1 \times 10^{10}\,\text{cm}^{-2}$, $\delta = 0.025$ (5%), $L = 500\,\mu\text{m}$, and $T = 300\,\text{K}$, respectively, unless otherwise specified. The corresponding critical tolerable parameters are $N_S^{\min} = 2.1 \times 10^{10}\,\text{cm}^{-2}$, $\delta^{\max} = 0.074$ (14.8%), and $L^{\min} = 170\,\mu\text{m}$.

Fig. 5 shows the threshold current densities for the main and the next modes, j_{th} and j_2, and the multimode generation threshold δj against normalized N_S (a), δ (b), and L (c). Fig. 38 shows the same quantities against T. The insets show the relative multimode generation threshold $\delta j/j_{\text{th}}$.

For the structures optimized to minimize j_{th}, Fig. 39 shows $\delta j/j_{\text{th}}$ and the minimum j_{th} against δ (a), L (b), and T (c). Each point on the curves corresponds to the specific structure optimized at the given δ (a), L (b) and T (c), respectively. For the optimized structures with $\delta = 0.025$ (5% dispersion) and 0.1 (20% dispersion), $\delta j/j_{\text{th}} \approx 21$ and 2%, respectively; the minimum j_{th} is 8 and 25 A/cm^2.

7.2.2. Relatively low temperatures

At relatively low temperatures (when nonequilibrium filling of QDs is the case below and at the lasing threshold), $\tau_{n,p}^{\text{esc}} \gg \tau_{\text{QD}}$ (and hence, equivalently, $j_{n,p}^{\text{esc}} \ll j_{\text{QD}}^{\text{sp}}$), $n_0(0)$ and $p_0(0)$ are $(1/\sigma_{n,p}v_{n,p}\tau_{\text{QD}})f_{n0}f_{p0}/(1 - f_{n0,p0})$, respectively [see (26)].

In this case, δj is controlled by the reciprocal of the spontaneous lifetime in QDs:

$$\delta j = 2(f_{n0} + f_{p0} - 1)\,\delta G\,\frac{f_{n0} + f_{p0} - f_{n0}f_{p0}}{2 - f_{n0} - f_{p0}}\frac{eN_S}{\tau_{\text{QD}}}. \tag{127}$$

The point is that the multimode generation threshold is controlled by the fastest process of the carrier escape from QDs. In the equilibrium case, this process is the thermally excited escape and hence the reciprocals of $\tau_{n,p}^{\text{esc}}$ enter into (122). In the nonequilibrium case, carrier leakage through the spontaneous recombination in QDs is the fastest process and hence the reciprocal of τ_{QD} enters into (127).

Were it not for the violation of the charge neutrality in QDs (which causes the slight temperature dependence of $f_{n0,p0}$), δj would be essentially temperature-independent at low T.

When passing from low to high temperatures, δj increases by the factor

$$\frac{2 - f_{n0} - f_{p0}}{f_{n0} + f_{p0} - f_{n0}f_{p0}} \frac{\tau_{QD}}{\tau_n^{esc}(1 - f_{n0}) + \tau_p^{esc}(1 - f_{p0})} \gg 1$$

where the high temperature values of $\tau_{n,p}^{esc}$ enter. At room temperature for the structure considered below, this factor is 24.

8. Conclusions

Theoretical analysis of the gain and the threshold current of a QD laser has been given which takes account of the line broadening caused by fluctuations in QD sizes. The following processes have been taken into consideration together with the main process of radiative recombination of carriers in QDs: band-to-band radiative recombination in the OCL, capture into QDs and thermally excited escape from QDs, and photoexcitation from QDs to continuum. Two regimes of QD filling by carriers, nonequilibrium and equilibrium, have been identified, depending on temperature, barrier heights and QD size. Critical tolerable parameters of the QD structure, at which lasing becomes impossible to attain, have been shown to exist. For an arbitrary QD size distribution, expressions for the threshold current density as a function of the RMS of relative QD size fluctuations, total losses, surface density of QDs, and thickness of the OCL have been obtained in an explicit form. The minimum threshold current density and optimum parameters of the structure (surface density of QDs and thickness of the OCL) have been calculated as universal functions of the main dimensionless parameter of the theory developed. This parameter is the ratio of the stimulated transition rate in QDs at the lasing threshold to the spontaneous transition rate in the OCL at the transparency threshold.

The electron and hole level occupancies in QDs have been obtained through the solution of the problem for the electrostatic field distribution across the junction. They were shown to differ from each other. As a result, the local neutrality is broken down in each QD. The key dimensionless parameters controlling the QD charge have been revealed. These parameters are governed by the surface density of QDs, thickness of the OCL, band offsets at the heteroboundaries, impurity concentrations in the cladding layers, quantized energy levels in QDs, and temperature. The lack of charge neutrality significantly effects the threshold current and its temperature dependence. The gain–current dependence of a laser has been calculated. The voltage dependences of the electron and hole level occupancies, gain and of the current have been obtained. The optimum surface density of QDs, minimizing the threshold current density, is distinctly higher than that calculated assuming the charge neutrality in QDs.

The detailed theoretical analysis of the temperature dependence of threshold current has been given. Temperature dependences of the current components associated with the radiative recombination in QDs and in the OCL have been calculated. Violation of the charge neutrality in QDs has been shown to give rise to the slight temperature dependence of the current component associated with the recombination in QDs. Increase of the temperature has been shown to suppress violation of the charge neutrality. The temperature T_d has been calculated (as a function of the parameters of the structure) at which the components of threshold current become equal to each other.

Temperature dependences of the optimum surface density of QDs and the optimum thickness of the OCL, minimizing the threshold current density, have been obtained. The lower bound for the set of curves for the temperature dependence of the threshold current has been found. This bound is the curve for the temperature dependence of the minimum threshold current density.

The characteristic temperature T_0 has been calculated considering carrier recombination in the OCL (barrier regions) and violation of the charge neutrality in QDs. The inclusion of violation of the charge neutrality has been shown to be critical for the correct calculation of T_0 at low T. The characteristic temperature has been shown to fall off profoundly with increasing temperature. A drastic decrease in T_0 occurs in passing from temperature conditions wherein the threshold current density is controlled by radiative recombination in QDs to temperature conditions wherein the threshold current density is controlled by radiative recombination in the OCL (i.e., by the thermoactivative leakage of carriers from QDs to the barrier regions). The dependences of T_0 on the RMS of relative QD size fluctuations, total losses and surface density of QDs have been obtained. The upper bound for the set of curves for the temperature dependence of T_0 has been found. This bound is the curve for T_0 calculated for the structures optimized with respect to both the surface density of QDs and the OCL thickness at each T.

Theoretical estimations presented here confirm the possibility of a significant reduction of the threshold currents and enhancement of T_0 of QD lasers as compared with the conventional QW lasers.

Detailed theoretical analysis of longitudinal spatial hole burning in QD lasers has been given. Unlike conventional semiconductor lasers, thermally excited escapes of carriers from QDs, rather than diffusion, have been shown to control the smoothing-out of the spatially nonuniform population inversion and multimode generation in QD lasers. The multimode generation threshold has been calculated as a function of structure parameters (surface density of QDs, QD size dispersion, and cavity length) and temperature. A decrease in the QD size dispersion has been shown not only to decrease the threshold current but to increase considerably the relative multimode generation threshold as well. Concurrent with the increase of threshold current, an increase of the multimode generation threshold has been shown to occur with a rise in temperature. Ways to optimize the QD laser, aimed at maximizing the multimode generation threshold, have been outlined.

Acknowledgements

The work was supported by the Russian Foundation for Basic Research and the Program "Physics of Solid State Nanostructures" of the Ministry of Industry and Science of the Russian Federation.

References

1. R. Dingle and C. H. Henry, "Quantum effects in heterostructure lasers", *U.S. Patent 3982207*, Sept. 21, 1976.
2. Y. Arakawa and H. Sakaki, "Multidimensional quantum well laser and temperature dependence of its threshold current", *Appl. Phys. Lett.* **40** (1982) 939-941.
3. D. Bimberg, M. Grundmann, and N. N. Ledentsov, *Quantum Dot Heterostructures*, John Wiley & Sons, Chichester, 1999, 328 p.
4. N. N. Ledentsov, V. M. Ustinov, A. Yu. Egorov, A. E. Zhukov, M. V. Maksimov, I. G. Tabatadze, and P. S. Kop'ev, "Optical properties of heterostructures with InGaAs-GaAs quantum clusters", *Semicond.* **28** (1994) 832-834.
5. N. Kirstädter, N. N. Ledentsov, M. Grundmann, D. Bimberg, V. M. Ustinov, S. S. Ruvimov, M. V. Maximov, P. S. Kop'ev, Zh. I. Alferov, U. Richter, P. Werner, U. Gösele, and J. Heydenreich, "Low threshold, large T_0 injection laser emission from (InGa)As quantum dots", *Electron. Lett.* **30** (1994) 1416-1417.
6. H. Hirayama, K. Matsunaga, M. Asada, and Y. Suematsu, "Lasing action of $Ga_{0.67}In_{0.33}As$/GaInAsP/InP tensile-strained quantum-box laser", *Electron. Lett.* **30** (1994) 142-143.
7. J. Temmyo, E. Kuramochi, M. Sugo, T. Nishiya, R. Nötzel, and T. Tamamura, "Strained InGaAs quantum disk laser with nanoscale active region fabricated with self-organisation on GaAs (311)B substrate", *Electron. Lett.* **31** (1995) 209-211.
8. R. Mirin, A. Gossard, and J. Bowers, "Room temperature lasing from InGaAs quantum dots", *Electron. Lett.* **32** (1996) 1732-1734.
9. D. Bimberg, N. Kirstaedter, N. N. Ledentsov, Zh. I. Alferov, P. S. Kop'ev, and V. M. Ustinov, "InGaAs–GaAs quantum-dot lasers", *IEEE J. Select. Topics Quantum Electron.* **3** (1997) 196-205.
10. N. N. Ledentsov, M. Grundmann, F. Heinrichsdorff, D. Bimberg, V. M. Ustinov, A. E. Zhukov, M. V. Maximov, Zh. I. Alferov, and J. A. Lott, "Quantum-Dot Heterostructure Lasers", *IEEE J. Select. Topics Quantum Electron.* **6** (2000) 439-451.
11. P. Bhattacharya, D. Klotzkin, O. Qasaimeh, W. Zhou, S. Krishna, and D. Zhu, "High-Speed Modulation and Switching Characteristics of In(Ga)As-Al(Ga)As Self-Organized Quantum-Dot Lasers", *IEEE J. Select. Topics Quantum Electron.* **6** (2000) 426-438.
12. L. Harris, D. J. Mowbray, M. S. Skolnick, M. Hopkinson, and G. Hill, "Emission spectra and mode structure of InAs/GaAs self-organized quantum dot lasers", *Appl. Phys. Lett.* **73** (1998) 969-971.
13. E. O'Reilly, A. Onishchenko, E. Avrutin, D. Bhattacharyya, and J. H. Marsh, "Longitudinal mode grouping in InGaAs/GaAs/AlGaAs quantum dot lasers: Origin and means of control", *Electron. Lett.* **34** (1998) 2035-2037.
14. P. M. Smowton, E. J. Johnston, S. V. Dewar, P. J. Hulyer, H. D. Summers, A. Patane, A. Polimeni, and M. Henini, "Spectral analysis of InGaAs/GaAs quantum-dot lasers", *Appl. Phys. Lett.* **75** (1999) 2169-2171.
15. D. L. Huffaker, G. Park, Z. Zou, O. B. Shchekin, and D. G. Deppe, "Continuous-Wave Low-Threshold Performance of 1.3-μm InGaAs-GaAs Quantum-Dot Lasers", *IEEE J. Select. Topics Quantum Electron.* **6** (2000) 452-461.
16. M. Sugawara, K. Mukai, Y. Nakata, K. Otsubo, and H. Ishilkawa, "Performance and Physics of Quantum-Dot Lasers with Self-Assembled Columnar-Shaped and 1.3-μm Emitting InGaAs Quantum Dots", *IEEE J. Select. Topics Quantum Electron.* **6** (2000) 462-474.
17. G. T. Liu, A. Stintz, H. Li, T. C. Newell, A. L. Gray, P. M. Varangis, K. J. Malloy, and L. F. Lester, "The Influence of Quantum-Well Composition on the Performance of Quantum Dot Lasers Using InAs/InGaAs Dots-in-a-Well (DWELL) Structures", *IEEE*

J. Quantum Electron. **36** (2000) 1272-1279.
18. V. P. Evtikhiev, I. V. Kudryashov, E. Yu. Kotel'nikov, V. E. Tokranov, A. N. Titkov, I. S. Tarasov, and Zh. I. Alferov, "Continuous stimulated emission at $T = 293$ K from separate-confinement heterostructure diode lasers with one layer of InAs quantum dots grown on vicinal GaAs(001) surfaces misoriented in the [010] direction in the active region", *Semicond.* **32** (1998) 1323-1327.
19. J. K. Kim, R. L. Naone, and L. A. Coldren, "Lateral Carrier Confinement in Miniature Lasers Using Quantum Dots", *IEEE J. Select. Topics Quantum Electron.* **6** (2000) 504-510.
20. M. Asada, Y. Miyamoto, and Y. Suematsu, "Gain and the Threshold of Three-Dimensional Quantum-Box Lasers", *IEEE J. Quantum Electron.* **22** (1986) 1915-1921.
21. K. J. Vahala, "Quantum box fabrication tolerance and size limits in semiconductors and their effect on optical gain", *IEEE J. Quantum Electron.* **24** (1988) 523-530.
22. Y. Miyamoto, Y. Miyake, M. Asada, and Y. Suematsu, "Threshold Current Density of GaInAsP/InP Quantum-Box Lasers", *IEEE J. Quantum Electron.* **25** (1989) 2001-2006.
23. L. V. Asryan and R. A. Suris, "Linewidth broadening and threshold current density of quantum-box laser", in *Proc. Int. Symp. "Nanostructures: Physics and Technology"*, St. Petersburg, Russia, June 1994, pp. 181-184.
24. R. A. Suris and L. V. Asryan, "Quantum-Dot Laser: Gain Spectrum Inhomogeneous Broadening and Threshold Current", in *Proc. SPIE's Int. Symp. PHOTONICS WEST'95*, San Jose, CA, USA, Feb. 1995, vol. 2399, pp. 433-444.
25. L. V. Asryan and R. A. Suris, "Inhomogeneous Line Broadening and the Threshold Current Density of a Semiconductor Quantum Dot Laser", *Semicond. Sci. Technol.* **11** (1996) 554-567.
26. L. V. Asryan and R. A. Suris, "Charge neutrality violation in quantum dot lasers", *IEEE J. Select. Topics Quantum Electron.* **3** (1997) 148-157.
27. L. V. Asryan and R. A. Suris, "Characteristic temperature of quantum dot laser", *Electron. Lett.* **33** (1997) 1871-1872.
28. L. V. Asryan and R. A. Suris, "Temperature dependence of the threshold current density of a quantum dot laser", *IEEE J. Quantum Electron.* **34** (1998) 841-850.
29. L. V. Asryan and R. A. Suris, "Spatial hole burning and multimode generation threshold in quantum-dot lasers", *Appl. Phys. Lett.* **74** (1999) 1215-1217.
30. L. V. Asryan and R. A. Suris, "Role of thermal ejection of carriers in the burning of spatial holes in quantum dot lasers", *Semicond.* **33** (1999) 981-984.
31. L. V. Asryan and R. A. Suris, "Longitudinal spatial hole burning in a quantum-dot laser", *IEEE J. Quantum Electron.* **36** (2000) 1151-1160.
32. L. V. Asryan and R. A. Suris, "Carrier photoexcitation from levels in quantum dots to states of the continuum during laser operation", *Semicond.* **35** (2001) 343-346.
33. N. N. Ledentsov, "Ordered arrays of quantum dots", in *Proc. 23rd Int. Conf. Phys. Semicond.*, vol. 1. Berlin, Germany, July 1996, pp. 19-26.
34. M. Grundmann and D. Bimberg, "Theory of random population for quantum dots", *Phys. Rev. B* **55** (1997) 9740-9745.
35. H. Benisty, C. M. Sotomayor-Torres, and C. Weisbuch, "Intrinsic mechanism for the poor luminescence properties of quantum-box systems", *Phys. Rev. B* **44** (1991) 10945-10948.
36. U. Bockelmann and G. Bastard, "Phonon scattering and energy relaxation in two-, one-, and zero-dimensional electron gases", *Phys. Rev. B* **42** (1990) 8947-8951.
37. U. Bockelmann and T. Egeler, "Electron relaxation in quantum dots by means of Auger processes", *Phys. Rev. B* **46** (1992) 15574-15577.
38. E. H. Perea, E. E. Mendez, and C. G. Fonstad, "Electroreflectance of indium gallium

arsenide phosphide lattice matched to indium phosphide", *Appl. Phys. Lett.* **36** (1980) 978-980.
39. S. Adachi, "Refractive indices of III-V compounds: Key properties of InGaAsP relevant to device design", *J. Appl. Phys.* **53** (1982) 5863-5869.
40. S. Adachi, "Material parameters of $In^{1-x}Ga^xAs^yP^{1-y}$ and related binaries", *J. Appl. Phys.* **53** (1982) 8775-8792.
41. G. P. Agrawal and N. K. Dutta, *Long-Wavelength Semiconductor Lasers*, Van Nostrand Reinhold Company, New York, 1986, 474 p.
42. D. Leonard, S. Fafard, K. Pond, Y. H. Zhang, J. L. Merz, and P. M. Petroff, "Structural and optical properties of self-assembled InGaAs quantum dots", *J. Vac. Sci. Technol. B* **12** (1994) 2516-2520.
43. M. Grundmann, R. Heitz, N. Ledentsov, O. Stier, D. Bimberg, V. M. Ustinov, P. S. Kop'ev, Zh. I. Alferov, S. S. Ruvimov, P. Werner, U. Gösele, and J. Heydenreich, "Electronic structure and energy relaxation in strained InAs/GaAs quantum pyramids", *Superlattices and Microstructures* **19** (1996) 81-95.
44. M. Sopanen, M. Taskinen, H. Lipsanen, and J. Ahopelto, "Visible luminescence from quantum dots induced by self-organized stressors", in *Proc. 23rd Int. Conf. Phys. Semicond.*, vol. 2. Berlin, Germany, July 1996, pp. 1409-1412.
45. J. I. Pankove, "Temperature dependence of emission efficiency and lasing threshold in laser diodes", *IEEE J. Quantum Electron.* **4** (1968) 119-122.
46. H. Statz, C. L. Tang, and J. M. Lavine, "Spectral output of semiconductor lasers", *J. Appl. Phys.* **35** (1964) 2581-2585.
47. M. A. Alam, "Effects of carrier transport on $L-I$ characteristics of QW lasers in the presence of spatial hole burning", *IEEE J. Quantum Electron.* **33** (1997) 1018-1024.
48. R. A. Suris and S. V. Shtofich, "Multifrequency stimulated emission from injection semiconductor lasers", *Soviet Phys. Semicond.* **16** (1982) 851-853.
49. R. A. Suris and S. V. Shtofich, "Role of impurities in the appearance of multifrequency emission from injection semiconductor lasers", *Soviet Phys. Semicond.* **17** (1983) 859-861.
50. L. V. Asryan, M. Grundmann, N. N. Ledentsov, O. Stier, R. A. Suris, and D. Bimberg, "Effect of excited-state transitions on the threshold characteristics of a quantum dot laser", *IEEE J. Quantum Electron.* **37** (2001) 418-425.
51. L. V. Asryan, M. Grundmann, N. N. Ledentsov, O. Stier, R. A. Suris, D. Bimberg, "Maximum modal gain of a self-assembled InAs/GaAs quantum-dot laser", *J. Appl. Phys.* **90** (2001) 1666-1668.
52. H. Hirayama, J. Yoshida, Y. Miyake, and M. Asada, "Carrier capture time and its effect on the efficiency of quantum-well lasers", *IEEE J. Quantum Electron.* **30** (1994) 54-62.
53. M. Sugawara, K. Mukai, and Y. Nakata, "Light emission spectra of columnar-shaped self-assembled InGaAs/GaAs quantum-dot lasers: Effect of homogeneous broadening of the optical gain on lasing characteristics", *Appl. Phys. Lett.* **74** (1999) 1561-1563.

APPLICATIONS OF QUANTUM DOTS IN SEMICONDUCTOR LASERS

Nikolai N. Ledentsov[*,a,b], Victor M. Ustinov[b], Dieter Bimberg[a], James A. Lott[c], and Zh. I. Alferov[b]

[a] Institut für Festkörperphysik, TU Berlin, Hardenbergstr. 36, D-10623 Berlin, Germany

[b] A.F.Ioffe Physical-Technical Institute, Politekhnicheskaya 26, 194021, St.Petersburg, Russia

[c] Air Force Institute of Technology, 2950 P Street B640, Wright-Patterson AFB, Ohio USA 45433

Quantum Dots (QD) provide unique opportunities to extend all the basic properties of heterostructure lasers and move further their applications. Practical fabrication of QD lasers became possible when techniques for self-organized growth allowed fabrication of dense and uniform arrays of narrow-gap nanodomains, coherently inserted in a semiconductor crystal matrix. Using of InAs QD lasers enabled significant improvement of device performance and extension of the spectral range on GaAs substrates to mainstream telecom wavelengths. Continuous wave 1.3 μm room-temperature output power of ~300 mW single mode for edge-emitters and of 1.2 mW multimode for vertical-cavity surface-emitting lasers are realized. Long operation lifetimes are manifested. The breakthrough became possible both due to development of self-organized growth and defect-reduction techniques in QD rechnology.

Keywords: quantum dot, heterostructure laser, VCSEL

1. INTRODUCTION

Nanoscale coherent insertions of narrow gap material in a single-crystalline matrix, or quantum dots (QDs) provide unique opportunities to modify and extend all the basic principles of heterostructure lasers and move further their applications. Since the first demonstration of photoumped and injection lasing in self-organized quantum dots in 1993 and 1994, respectively, these devices are under intense studies all over the world. Originally the idea to "exploit quantum effects in heterostructure semiconductor lasers to produce wavelength tunability" and achieve "lower lasing thresholds" via "the change in the density of states which results from reducing number of translational degrees of freedom of the carriers" was introduced in 1976 [1]. The possibility to achieve a singularity in the density of states (see Fig.1) for structures with size quantization in more than one direction has been also considered in Ref.1. However, lasers using structures with carrier confinement in two ("quantum wire") or all three ("quantum dot") directions failed in practical realization due to the lack of appropriate technology at that time. The predominant motivation of using of quantum wires and quantum dots in lasers [1] over the thick-layer or quantum well devices [1,2] is the theoretically predicted advantages of a gain medium with zero-dimensional density of states, such as low threshold current density, J_{th}, high characteristic temperature, T_0 and high material and differential gain. However, lasers using structures with carrier confinement in two ("quantum wire") or all three ("quantum dot") directions failed in practical realization due to the lack of appropriate technology at that time. The real breakthrough occurred when new class of heterostructures - self-organized quantum dots was applied to lasers [3,4]. From the engineering point of view QDs can offer an advantage of reduced transport of nonequilibrium carriers out of the injection region and/or suppress nonradiative recombination at laser facets. Particularly strong interest attract in this case high power QD lasers and small area vertical-cavity surface-emitting lasers where facet overheating and carrier spreading play important roles, respectively. High power lasers [2] play a crucial role in many areas, from material processing to telecom applications. A possibility of high-power long wavelength operation is necessary for telecom pumps at 1.48 µm, high power single transverse mode telecom lasers of the 1.3 µm and 1.55 µm ranges, eye-safe long-wavelength lidars and long-wavelength high-power phase-locked VCSEL arrays for laser scanning systems. However, as opposite to GaAs-based quantum well devices emitting near 1 µm, where CW output powers of about 10 W and CW power densities per facet area of 30-40 mW/cm^2 are obtained for 100 µm wide ridge stripe lasers [1], the advances in InP-based long-wavelength lasers are less significant. This motivates the attempts to bring the GaAs-based laser technology to the long-wavelength (1.3-1.55 µm) wavelength range. Here, despite of the late start in 1998, long-wavelength QD laser technology has gained significant advances.

2. GROWTH OF QUANTUM DOTS

2.1. General

To fabricate a quantum dot laser, one needs to have a dense array of quantum dots uniform in size and shape and free from undesirable defects. The latter point is the crucial and the first QD laser based on InGaAs-GaAs QDs was using defect-free islands of relatively small size emitting near 1 µm [3,4].

The technique of QD fabrication that was later applied for current-injection GaAs-based QD lasers [3,4] employs on self-organized growth of uniform nanometer-scale islands. It is known that a layer of a material having a lattice constant different from that of the substrate, after some critical thickness is deposited, may spontaneously transform to an array of *three-dimensional* islands [5]. More recently it was shown on an example of InGaAs QDs on a GaAs surface, that there can exist a range of deposition parameters, where the islands are small (~10 nm), have a similar size and shape and form dense arrays [4, 6]. Due to the *strain-induced renormalization* of the surface energy of the facets, an array of *equisized and equishaped* 3D islands can represent a *stable state* of the system [7] (so called size-limited island growth, SLIG). Interaction of the islands via the substrate makes also their *lateral ordering* favorable. If strained InGaAs islands are covered with a thin GaAs layer, the islands in a second sheet are formed over the dots in the first sheet resulting in a *three-dimensional ordered array* of QDs, either isolated or strongly *vertically-coupled* [8]. The size and the shape of the SLIG InGaAs islands can be changed by replacing InAs by InGaAs or InGaAlAs and by changing the deposition mode. The wavelength range of the SLIG InAs islands is around 1.2-1.26 µm depending on the substrate temperature.

2.2. Growth fundamentals of QDs

An increase in the substrate temperature increases the density of adatoms on the surface and decreases the total number of atoms arranged in QDs.

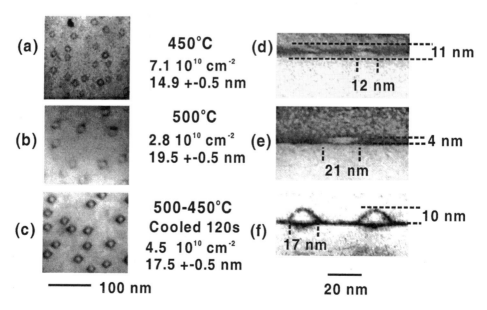

Fig. 1 Plan-view (a-c) and cross-section (d-e) TEM images of InAs QDs.

Reversible and irreversible phenomena in Stranski-Krastanow (SK) growth of strained three-dimensional (3D) InAs islands on GaAs(001) surface were studied in [9]. Transmission electron microscopy combined with photoluminescence (PL) spectroscopy has revealed the following. Increase in the substrate temperature during the InAs deposition results in a *decrease* in the island density, in an *increase* in their lateral size, and in a *decrease* in the island volume. The SLIG island's density and the lateral size as well as PL spectra do not depend on the growth interruption time in the range of 10-40 s before the islands are capped with GaAs. If the substrate temperature is reduced *after* the InAs deposition, the density of islands increases, the average volume of the single island increases, while their lateral sizes undergo shrinkage. This effect is illustrated by Fig. 1. The dots ramped from 500 to 450°C within 90-120 s and covered at the final temperature demonstrate bright PL emission at room temperature near 1.3 µm [10].

2.3. Long-wavelength emission in structures with vertically-coupled QDs

Another way to produce GaAs-based QDs with emission tunable in a wide spectral range is to grow vertically-coupled QDs [8, 11]. Electronic coupling causes a significant longwavelength shift of the PL emission with increase in the number of stacks, even in the case where very small islands are used as stacking objects. Strongly electronically coupled quantum dots, forming QD superlattices in vertical direction, or superdots, have been demonstrated also in MOCVD [12] (see Fig. 2). Using this approach it is possible to fabricate QD structures emitting in the range 1.3-1.44 µm.

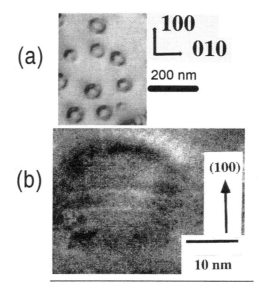

Fig. 2. Plan-view (a) and cross-section (b) TEM images of QDs, formed by multi-cycle InGaAs-GaAs deposition with spacer layer thickness much smaller than the height of single QD.

2.4. Long-wavelength emission in QDs formed by activated alloy phase separation

Another way to achieve long-wavelength emission from GaAs-based QDs is to use the effect of activated alloy phase separation decomposition. It is based on the controlled increase of the volume of the strained island by alloy overgrowth and was demonstrated for InAs QDs on GaAs substrates covered by (In,Ga)As alloy layer [13]. Initially, an array of small coherent InAs stressors is formed. If these InAs islands formed by ~2 ML InAs deposition are covered by GaAs, the resulting QDs have a high density, but a rather small lateral size and height and may be revealed only in a strong beam TEM conditions (Fig.3, left TEM image). Once the dots are covered by InGaAs alloy, it is energetically favorable for InAs molecules to nucleate at the elastically-relaxed islands, where the lattice parameter is close to that in InAs.

This gives a possibility to create dense array of large coherent QDs emitting up to 1.32 µm at room temperature (see Fig.4 right polar TEM image).

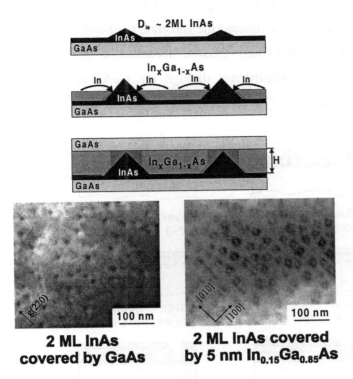

Fig.3 Large InAs QDs formed by activated alloy phase separation

2.5. Defects in long-wavelength QDs and defect reduction techniques

Formation of coherent InAs islands is accompanied in many cases by formation of dislocated clusters. As these clusters attract Indium atoms more effectively as compared to elastically-relaxed islands, the size of the dislocated objects is larger. This enables us to select the islands on the basis of their height. In these technique the InGaAs islands are covered by a thin GaAs layer with nominal thickness sufficient to overgrow smaller coherent islands.

Large or dislocated islands remain uncovered. The dislocated islands have the strongest lattice mismatch with GaAs and cause the strongest repulsion of the GaAs upon overgrowth. At the stage when large and/or dislocated islands are not covered the substrate temperature is increased (see Fig.4). During this procedure InGaAs, accumulated in dislocated clusters evaporates and (or) redistributes through the GaAs surface forming the second wetting layer.

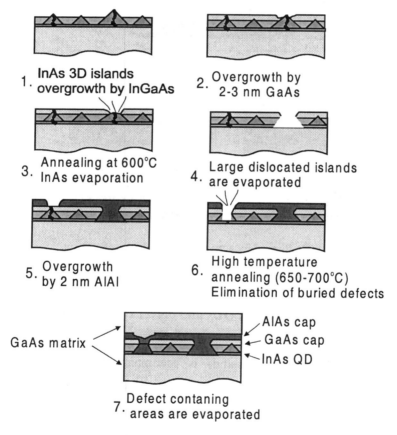

Fig.4. Defect-reduction technique

The procedure can be further improved by depositing a thin AlAs layer, which does not nucleate at the strain-relaxed regions followed by high-temperature evaporation of dislocated regions. This approach led to the formation of defect-free long-wavelength GaAs-based QDs using MOCVD, their successful stacking and realization of room-temperature gain and stimulated emission at wavelengths up to 1.38 μm [14].

In Fig.5 we show plan-view (only the upper sheet of QDs is trapped by the TEM foil in the latter case) and cross-section TEM images of MOCVD InGaAs-GaAs QDs formed by InGaAs deposition. The structure grown without the annealing step (Fig.5a,b) is completely dislocated after the stacking. Coherent dots persist only in the lower sheet, while in the upper sheet all InGaAs exceeding the thickness of the wetting layer is absorbed by the dislocation regions.

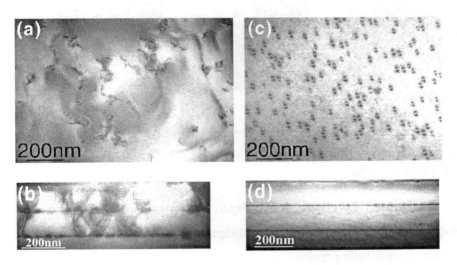

Fig.5 Plan-view (a,c) and cross-section (b,d) TEM images of MOCVD QDs grown without (a,b) and with (c,d) annealing step.

2.6. Long-wavelength emission from wavelength InGaAsN insertions in a GaAs matrix

InGaAsN insertions in a GaAs matrix represent an example of the alloy, that effectively decomposes to QDs both by formation of 3D islands and by alloy phase separation. Alloy phase separation dominates at lower In and higher N content, while formation of islands dominates at higher In content. Already the structures emitting at wavelength near 1.2 μm represent QDs of relatively large size, and the structures emitting closer to 1.3 μm demonstrate well-resolved pyramidal shape of the self-organized QDs in cross-section TEM images.

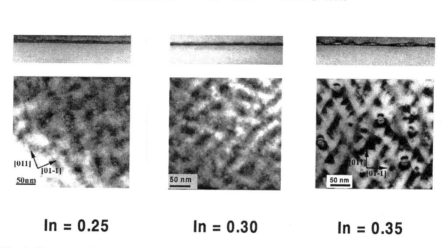

Fig.6. Cross-section (top) and plan view TEM images of InGaAsN insertions with different indium content.

For lower In content (see Fig.6) the plan-view TEM contrast corresponds to the network of In-rich elongated domains surrounded by As-rich strain-compensating regions.

3. INJECTION LASERS BASED ON QUANTUM DOTS

3.1. QD lasers

At low temperatures QD lasers based on QDs demonstrated good device characteristics practically from the time of their first realization in 1993-1994. On the other hand already in 1994 [4] it was recognized that the main obstacle for QDHS laser operation at elevated temperatures is related to temperature-induced escape of carriers from QDs. Several approaches were proposed to improve the performance (see [15] and references therein):

(i) the increase of the density of QDs by stacking of QDs,
(ii) the insertion of QDs in a QW,
(iii) the use of matrix material having a higher band-gap energy,
(iv) the increase of the density of QDs by using the concept of "seeding" of QDs.

Using the concept of electronically coupled in vertical direction InGaAs QDs in a GaAlAs matrix permitted us to reduce the threshold current density further to 63 A/cm^2 [11] and realize high-power continuous wave (CW) lasing [15].

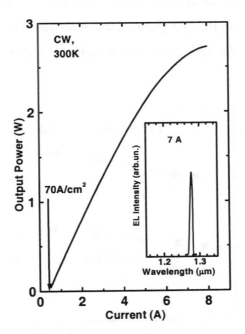

Fig.7. High-power operation of a QD laser

Further optimization led to an increase of the output power up to 3.5-4 W CW [15] for 100-µm-stripe width. A quantum efficiency of 95% and a wall-plug efficiency of 51% were obtained. However, the most crucial advantage of QDs for edge-emitters and vertical-cavity surface emitting lasers (VCSELs) is related to the possibility to cover the strategically important spectral ranges of 1.3 and 1.55 µm using GaAs substrates. As cost-effective high reflectivity monolithic Bragg reflectors and developed oxide technology are available for GaAs substrates, this is of key importance.

QD lasing at 1.3 µm on GaAs substrate was first reported in [16]. The best results in high-power operation are currently obtained using stacked QDs fabricated by activated alloy phase separation [13]. CW operation at room temperature (RT) [17] up to 2.8 W is achieved (see Fig. 7). For a stripe width of 100 µm and cavity lengths (L) exceeding 1200-1600 µm (uncoated) it was found that the threshold current density was only weakly dependent on the cavity length being about 70-90 A/cm^2. The lasing wavelength is near 1.3 µm and corresponds to the QD ground state emission. The estimated maximum modal gain for the QD ground state lasing is about 14 cm^{-1}, and can be

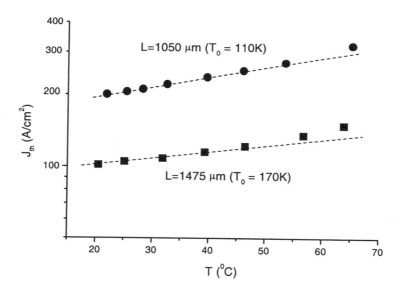

Fig.8. High temperature operation of a long-wavelength QD laser. Five-fold stacked QDs are used.

increased up to 35 cm^{-1} for 10-fold-stacked QD active regions. The internal losses derived from the slope of the inverse differential efficiency versus cavity length are as low as 1.5 cm^{-1}. QD lasers demonstrate improved temperature stability of the threshold current [13], and in the vicinity of room temperature the characteristic temperature approaches 170 K (see Fig. 8).

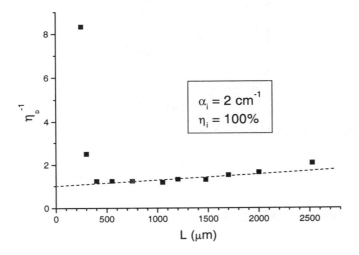

Fig. 9 High temperature dependence of differential efficiency. Five-fold stacked QDs are used.

Differential efficiency as high as 84% has been realized for 1.1-mm cavity (see Fig. 9) lengths (uncoated). The lasing wavelength approaches 1.3 μm at highest drive currents. Narrow 7 μm-wide stripes were also fabricated. Low threshold single-transverse-mode kink-free operation up to 200 mW was demonstrated for uncoated facets and 2 mm-long cavity [18] (see also Fig. 10).

For 1.25 μm QD lasers Liu et al. [19] demonstrated the reduction of the threshold current density to 26 A/cm^2 in 7 mm-long cavities (uncoated), while keeping sufficiently high differential efficiencies (31%).

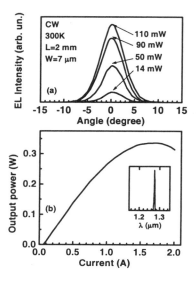

Fig. 10. High-power kink-free operation of long-wavelength QD laser.

3.2. Long wavelength QD VCSELs

Another important issue in the realization of long-wavelength GaAs-based QDs is the GaAs-based VCSEL emitting at 1.3 µm. It is known that arrays of VCSELs allow a possibility to reach up to 1 W CW operation while having a very low beam divergence. Recently, 1.3 µm VCSELs using QDs formed by activated alloy phase separation have been fabricated [20-22]. The structures are grown directly on GaAs substrates and when fabricated include selectively oxidized Al(Ga)O current apertures, intracavity metal contacts, and Al(Ga)O/GaAs distributed Bragg reflectors (DBRs). Devices operate at room temperature and above with threshold currents below 2 mA and differential slope efficiencies of about 40 percent in the CW mode [21]. In Fig. 11 we show light-current charactersitics of the device. The output power at 25°C heat sink temperature is 0.65 mW at 10-15% wall-plug efficiency. Preliminary degradation tests demonstrate sufficiently long operation lifetime of the device.

From direct observations of relaxation oscillations, cut-off frequencies larger than 10 GHz have been determined for edge-emitting QD-laser with 10-fold stacked QDs emitting at 1.27 µm [23]. It is clear from these results that QD lasers are practically useful for applications in optical transmitters with direct modulation.

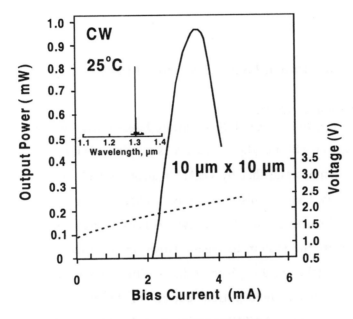

Fig. 11. CW operation of long-wavelength QD VCSEL

4. COMPARISON OF THE PERFORMANCE OF LONG-WAVELENGTH LASERS OBTAINED USING DIFFERENT TECHNIQUES

The emergence of semiconductor quantum dot technology promises to provide lasers with enhanced performance in all areas. QD technology can be applied to both edge emitters and VCSELs at both long communication wavelengths. In addition mature, and as compared to InP, inexpensive GaAs technology (availability of 6 inch wafers) might be used for production. A common technology for the present by far largest part of the semiconductor world laser market would be created. The consequent reduction of production cost for datacom lasers would be enormous. This first of a series of reports will review the published work of the world's leading research groups in this area up to 8/2000 and comment on those technical developments which could significantly impact business.

4.1. Competitors

4.1.1. NEC Corporation, Japan (Gas Source MBE)

GaAsSb insertions in GaAs.

Edge-emitting lasers: Edge-emitting lasers operating near 1.3 µm were realized using GaAsSb insertions in a GaAs matrix grown by gas-source MBE. The authors demonstrate that by GaAsSb quantum-wells embedded in GaAs layers, long-wavelength emission can be realized. According to the authors GaAsSb-GaAs structures are of weak type-II nature, and placing of GaAsSb in AlGaAs converts the system to type-I. However, no time-resolved studies to confirm this conclusion are given. Edge emitting lasers at 1.27-1.3 µm are demonstrated with threshold current density of 450-900 A/cm^2, respectively (cavity length 2.2 mm). VCSELs emitting at 1.2 µm are demonstrated with 3.3 kA/cm^2 [Nishi-SD]. More recently 800 A/cm^2 at 2 mm cavity length for 3-fold stacked GaAsSb insertions are reported. For an external loss of 30 cm^{-1} the threshold current density is 3 kA/cm^2 [24].

VCSELs: No 1.3 µm VCSEL is reported. 1.23 µm VCSEL lasing from double GaAsSb insertions in GaAs is reported. GaAs-AlAs doped DBRs are used. For 6 µm aperture the threshold curent was 0.7 mA (1994 A/cm^2). Maximum differential efficiency was 20% and maximum power 0.1 mW at RT [25].

4.1.2. Hitachi (MBE)

Edge-emitters: InGaAsN. 1.309 μm lasing at RT (800 μm x 2 μm, uncoated) is achieved. CW, 0.16W/A (Differential efficiency is 17%). J_{th}=6750 A/cm^2 [26]. Very high T_0 up to 214K (pulsed) is reported [27], but the threshold current density at 20°C (1.304 μm) was already 2.86 kA/cm2 (cavity length 800 μm).

VCSELs: InGaAsN. No VCSELs operating close to 1.3 μm are reported. 1.19 μm pulsed GaAs-based VCSEL is demonstrated [28].

4.1.3. University of Texas (MBE)

Edge emitters: InGaAs QDs 1.3 μm pulsed lasing at 300K with 150A/cm^2 threshold current density is realized for 5.54 mm cavity length (uncoated). [29]. CW lasing at 1.33 μm at 300K is realized in a structure with coated facets (HR/HR). J_{th}=19A/cm^2 (L=1139 μm, HR/HR, oxide-confined stripe). However, the external differential quantum efficiency was only about 2% at threshold and was decreasing with current. 290 μW output power at 24 mA was obtained [30]. T_0 in the CW regime near 300K is about 20-25K for 5040 μm cavity length (as cleaved) [31].

VCSELs: No VCSELs close to the 1.3 μm range. 1.15 μm oxide-confined VCSEL using InGaAs QDs is demonstrated [32].

4.1.4. Infineon (in cooperation with Ioffe institute)

Edge-emitters (InGaAsN-GaAs, plasma-source MBE) 1.294 μm edge-emitting laser with J_{th}=400A/cm^2 at 10°C (L=1120 μm, W=100 μm) is demonstrated for single InGaAsN insertion. For short cavity lengths (external losses around 35 cm^{-2}), J_{th} is increases to 1.1 kA/cm^2. Maximum CW output power is 2.4 W. Maximum differential efficiency is 63% [33]. Recently, using active cooling fixing the „on-chip" temperature 8W CW was measured. Output power for 10°C heat sink temperature was 4W CW [34].

VCSELs: VCSELs at 1.29 µm are realized with J_{th} around 10 kA/cm^2. CW output power is 1 mW at 10°C heat sink temperature.

4.1.5. Fujitsu (MBE)

Edge emitters (InAs-InGaAs QDs): 1.31 µm lasing is realized. Internal losses are 1.2 cm^{-1}. The lasing occurs in structures with highly reflective facet coatings. No data on output power, or differential efficiency is available is given. T_0=82K in a range 10-30°C (CW) as measured for a cavity length of 1800 µm with HR/HR coatings. No data for practically relevant devices is given [35].

VCSELs: No VCSELs are reported for 1.3 µm range.

4.1.6. Princeton University (MBE)

Edge emitters: InGaAsN. Lasing at 1.302 µm at RT is reported in a pulsed mode, J_{th} = 1.3 kA/cm^2 for 3 mm long cavity length (8 µm wide ridge stripe) and 1.5 kA/cm^2 at 1.1 mm cavity length. Internal quantum efficiency is 71%. Internal losses are 3 cm^{-1}. Maximum differential efficiency is 48% [36].

VCSELs: No VCSELs are reported.

4.1.7. University of New Mexico (MBE)

Edge-emitters: InAs QDs. 1.293 µm pulsed lasing is reported for 7.8 mm-long cavity with J_{th}=42 A/cm^2. Output power is about 50 mW. Differential quantum efficiency is ~20%. T_0 is not reported [37].

VCSELs: No VCSELs are reported

4.1.8. Sandia (MBE)

Edge emitters: GaAsSb-GaAs GaAsSb single-quantum-well lasers are grown on GaAs substrates. Room temperature pulsed emission at 1.275 µm in a 1250-µm-long device is achieved. Minimum threshold current density of 535 A/cm^2 was measured in 2000-µm-long devices. The measured internal losses are 2-5 cm^{-1}, the internal quantum efficiency is 30%-38% and the characteristic temperatures T_0 is 67°C-77 °C [38].

VCSELs (InGaAsN) 1.294 µm lasing with J_{th}~9.6 kA/cm^2 is achieved. N-doped AlAs-GaAs DBRs are used. Top N-doped-DBR is connected via Esaki junction to the p-injection layer. Operation voltage is 4.5 V, maximum output power 60 µW CW. Slope quantum efficiency is ~2% [39].

4.1.9. Ioffe (in cooperation with TU Berlin)

Edge-emitters: InAs-InGaAs QDs. 1.28 µm lasing with 3W CW is realized. The devices are fabricated in the ridge stripe geometry (width 200 µm), 1.9 mm cavity length, uncoated. J_{th}~90A/cm^2. Three-fold stacked InAs-InGaAs QDs grown by MBE are used [17]. Maximum differential efficiency was 84%. Saturation gain is about 20 cm^{-1} For 5-fold-stacked QDs. For 10-fold stacked QDs it can be increased to 35 cm^{-1} at a current density of 1.5 kA/cm^2 at 300K [15]. T_0 is about 170K in the temperature range 270-340K (1.45 mm cavity length, uncoated). Single transverse mode operation up to 200-300 mW CW is realized for 7-µm wide mesas (300K) in the CW regime [18].

VCSELs:InAs QDs (in cooperation with Wright-Patterson Air Force Base, Ohio, USA). 1.304 µm lasing is realized in the pulsed regime at 300K in a structure with intracavity contacts using fully-oxidized AlO-GaAs DBRs. Threshold currents are about 1 mA for 8 µm aperture. Maximum power is 1.2 mW CW. Maximum differential efficiency 46% pulsed [20, 21] and beyond 90% CW due to high negative T_0 [22]. Operation lifetimes in excess of 5000 hrs CW without degradation are demonstrated. Maximum wall-plug efficiency is 20%.

4.1.10. TU Berlin

Edge-emitters. TU Berlin activities on MOCVD-grown concentrated on high-power QD lasers emitting in the 1100-1150 nm range for pumping Tm^{3+} doped fluoride fiber lasers. Maximum total output pulsed power of 3.7 W is realized. Extremely low loss lasers (1.5 cm^{-1}), large internal quantum efficiency of 98%, and transparency current of 18 A/cm^2 (6A/cm^2 per QD stack) are demonstrated [41].

4.2. Comparison of process technologies, material systems and device performance in the work of different authors

There exists a problem to draw conclusions on the optimum process from different authors, some of them have a clear advantage in some particular device parameter, but a clear disadvantage in other characteristics. E.g. NEC reported high temperature stability of the threshold current of 1.3 µm (T_0=215K), but at the same time J_{th} is very high (>2.5 kA/cm^2) and the operation is pulsed. Infineon report 270A/cm^2, but T_0 is 60-70K, Fujitsu reports T_0 of 85K, but they use high-reflection coatings on both facets which makes the device not practical, etc. Concerning the edge emitters in the 1.3 µm range, comparable best results are obtained by Ioffe Institute (in cooperation with TUB) using InAs QDs and Infineon (in cooperation with Ioffe institute) using InGaAsN. Infineon-Ioffe team reported 4W CW for heatsink temperature of 10°C and Ioffe-TUB team reported 2.8 W CW for 17°C heatsink temperature.

For VCSELs the same teams are mostly competing. Ioffe Institute in cooperation with TUB and AFIT (USA) reported 1.3 µm VCSEL with I_{th}<1.2 mA for 8 µm square aperture (CW, 25°C heat sink temperature). J_{th}<2 kA/cm^2. Differential quantum efficiency is about 40% and maximum power is ~1.2 mW. Infineon has 1 mW CW operation at 10°C with J_{th}<2 kA/cm^2. Sandia has 2% differential quantum efficiency, J_{th}=9.6 kA/cm^2 and 1.294 µm emission wavelength (0.06 mW maximum power).

The comparison of the most important device parameters obtained by different teams and using different approaches is given in the Table.

Table.
Relative comparison of results of different teams.

Company, technique	λ µm	J_{th}, RT A/cm^2	J_{tr}, RT A/cm^2	α_I cm^{-1}	η_{int} %	η_{diff} %	T_0 K	Comments
Ioffe InAs QDsMBE Edge emitter	1.26 - 1.28	70 (L=2 mm)	<10	1.5	60	56	170	5-fold-stacked QDs 75-84% diff. eff.
VCSEL	1.28 - 1.31	~3000	-	-	-	40	-	3-fold-stacked QDs, CW 0.65 mW(25°C)
Sandia GaAsSb MBE Edge-emitters	1.28	535 (L=2 mm)	134	2-5	30-38	~15	67-77	Pulsed, single insertion
VCSEL-MBE InGaAsN	1.29	9600	-	-	-	2	-	0.06 mW CW,
NEC GaAsSb GS MBE Edge-emitter	1.27 - 1.30	470/1.27µm 770/1.3µm (L=2 mm)	560	-	-	20 (1.27 µm)	65 (for 1.27µm)	3-fold stacked GaAsSb insertions pulsed.
VCSEL	1.23	1994				20		0.1 mW..
Hitachi MBE InGaAsN Edge emitter	1.3-1.31	2860/1.305 6750/1.309 (L=0.8 mm)	-	-	-	17 (1.30 9 µm)	214 /1.304	Pulsed
VCSEL	only 1.15 µm							
University of Texas MBE InGaAs QDs Edge-emitter	1.33 µm	150 (5.54 mm) 19 (HR/HR)	-	-	-	2	20-25	Pulsed, 290 µW at 24 mA
VCSELs	only 1.15							

	μm							
University of New Mexico InAs QDs MBE Edge-emitters	1.293	42 (L=7.8 mm)	13	1	-	20	-	Pulsed
VCSEL	No							
Princeton University InGaAsN MBE Edge emitter	1.302	1300 (L=3 mm)	~1100	3	60	48	150	Pulsed CW 125 mW
VCSEL	No							
Infenion-Ioffe InGaAsN Edge emitter	1.300	270 (L=3.2 mm)	110	4	-	40-61	75	CW 4.2 W at 10°C
VCSEL	1.29	10000						1 mW CW at 10°C
Fujitsu InAs QD MBE	1.31	300 (HR/HR)	-	1.2		5	85 (HR/HR)	0.8 mW CW
VCSEL	No							

5. CONCLUSION

The progress in the field of QD lasers is remarkably fast and it is accelerating further. For the most important, commercially relevant applications near 1.3 µm QD lasers demonstrate a very high potential and already superb properties as compared to InGaAsN based devices, even the work in the 1.3 µm direction started later (1998 versus 1995). Edge-emitters near 1.3 µm of device quality are realised by Ioffe-TUB teams (QDs) and by Infineon-Ioffe teams. Additionally, Ioffe team claims that InGaAsN structures clearly represent QD structures, according to their TEM data. As it is clear from the comparison of the results of different teams, intentionally created QDs already overrun other approaches.

ACKNOWLEDGEMENTS

This work is supported by NonOp CC, INTAS, and Volkswagen Foundation. N.N.L. is supported by the DAAD Guest Professorship programme. Helpful discussions with A.Yu. Egorov, A.R. Kovsh, I.L. Krestnikov, M.V. Maximov, N.A. Maleev, N. Grote, C. Ribbat, R. Sellin, A.F. Tsatsul'nikov, B.V. Volovik, and A.E. Zhukov are gratefully acknowledged.

REFERENCES

1. R. Dingle and C.H. Henry, "Quantum effects in heterostructure lasers" U.S. Patent No. 3982207, 21, September, 1976.
2. D.A. Livshits, A.Yu. Egorov, I.V. Kochnev, V.A. Kapitonov, V.M. Lantratov, N.N. Ledentsov, T.A. Nalet, I.S. Tarasov, "Record power characteristics of InGaAs/AlGaAs/GaAs heterostructures" Fiz. I Tekh. Poluprovodn. (Semiconductors) 35, 380-384 (2001).
3. N.N. Ledentsov, V.M. Ustinov, A.Yu. Egorov, A.E. Zhukov, M.V. Maximov, I.G. Tabatadze, P.S. Kop'ev, "Optical properties of heterostructures with InGaAs-GaAs quantum clusters" *Fiz. I Tekh. Poluprovodn.* Vol. 28, pp. 1484-1488 - *Semiconductors* Vol. 28, pp. 832-834 (1994) (Submitted December 29, 1993).N.N. Ledentsov, M. Grundmann, N. Kirstaedter, J. Christen, R. Heitz, J. Böhrer, F. Heinrichsdorff, D. Bimberg, S.S. Ruvimov, P. Werner, U. Richter, U. Gösele, J. Heydenreich, V.M. Ustinov, A.Yu. Egorov, M.V. Maximov, P.S. Kop'ev and Zh.I. Alferov "Luminescence and structural properties of (In,Ga)As - GaAs quantum dots" *Proceedings of the 22nd International Conference on the Physics of Semiconductors*, Vancouver, Canada, 1994, Ed. D.J. Lockwood (World Scientific, Singapore, 1995), p.1855-1858, vol.3.
4. N.Kirstaedter, N.N.Ledentsov, M.Grundmann, D.Bimberg, V.M.Ustinov, S.S.Ruvimov, M.V.Maximov, P.S.Kop'ev, Zh.I.Alferov, U.Richter, P.Werner, U.Gosele J.Heydenreich, "Low threshold, large T_0 injection laser emission from (InGa)As quantum dots" *Electron. Lett.* Vol. 30, pp. 1416-1418 (1994).
5. L. Goldstein, F. Glas, J.Y. Marzin, M.N. Charasse, G. Leroux, "Growth by molecular-beam epitaxy and characterization of InAs/GaAs strained-layer superlattices." Appl. Phys. Lett., Vol.47, pp.1099-1101 (1985).
6. D. Leonard, M. Krishnamurthy, C.M. Reaves, S.P. Denbaars and P.M. Petroff, "Direct formation of quantum-sized dots from uniform coherent islands of InGaAs on GaAs surfaces" Appl. Phys. Lett., Vol. 63, pp. 3203-3205 (1993).
7. V.A. Shchukin, N.N. Ledentsov, P.S. Kop'ev, and D. Bimberg, "Spontaneous ordering of arrays of coherent strained islan ds" *Phys. Rev. Lett.* Vol. 75, 2968-2972 (1995).

8. N.N. Ledentsov, M. Grundmann, N. Kirstaedter, O. Schmidt, R. Heitz, J. Böhrer, D. Bimberg, V.M. Ustinov, V.A. Shchukin, P.S. Kop'ev, Zh.I. Alferov, S.S. Ruvimov, A.O. Kosogov, P. Werner, U. Richter, U. Gösele and J. Heydenreich, "Ordered Arrays of Quantum Dots: Formation, Electronic Spectra, Relaxation Phenomena, Lasing" *Proceedings MSS7*, 1995, *Solid State Electronics* Vol. **40**, 785-798 (1996).

9. V.A. Shchukin, N.N. Ledentsov, V.M. Ustinov, Yu.G Musikhin, B.V Volovik, A. Schliwa, O. Stier, R. Heitz, and D. Bimberg, „New tools to control morphology of self-organized quantum dot nanostructures" 2000 Spring Meeting of the Materials Research Society. April 24 – 28, 2000, San Francisco, USA. In: *Morphological and Compositional Evolution of Heteroepitaxial Semiconductor Thin*. Ed. By A.L. Barabási, E. Jones, and J. Mirecki Millunchick. Mat. Res. Soc. Symp. Proc. V. **618**. Pittsburgh, USA. 2000.

10. N.N. Ledentsov, V.A. Shchukin, D. Bimberg, V.M. Ustinov, N.A. Cherkashin, Yu.G. Musikhin, B.V. Volovik, G.E. Cirlin, and Zh.I. Alferov "Reversibility of the Island Shape, Volume and Density in Stranski - Krastanow Growth", to be published.

11. N.N. Ledentsov, V.A. Shchukin, M. Grundmann, N. Kirstaedter, J. Böhrer, O. Schmidt, D. Bimberg, S.V. Zaitsev, V.M. Ustinov, A.E. Zhukov, P.S. Kop'ev, Zh.I. Alferov, A.O. Kosogov, S.S. Ruvimov, P. Werner, U. Gösele and J. Heydenreich, "Direct formation of vertically coupled quantum dots in Stranski-Krastanow growth" *Phys. Rev. B* Vol. **54**, pp. 8743-4751 (1996).

12. N.N. Ledentsov, J. Böhrer, D. Bimberg, I.V. Kochnev, M.V. Maximov, P.S. Kop'ev, Zh.I. Alferov, A.O. Kosogov, S.S. Ruvimov, P.Werner and U. Gösele "Formation of coherent superdots using metal-organic chemical vapor deposition" *Appl. Phys. Lett.* 69, 1095 (1996)

13. Yu.M. Shernyakov, D.A. Bedarev, E.Yu. Kondrat'eva, P.S. Kop'ev, A.R. Kovsh, N.A. Maleev, M.V. Maximov, V.M. Ustinov, B.V. Volovik, A.E. Zhukov, Zh.I. Alferov, N.N. Ledentsov, D. Bimberg, "1.3 µm GaAs-based laser using quantum dots obtained by activated spinodal decomposition", *Electronics Letters* Vol. 35, 898-900 (1999); M.V. Maximov, A.F. Tsatsul'nikov, B.V. Volovik, D.S. Sizov, Yu.M. Shernyakov, I.N. Kaiander, A.E. Zhukov, A.R. Kovsh, S.S. Mikhrin, V.M. Ustinov, Zh.I. Alferov, R. Heitz,

V.A. Shchukin, N.N. Ledentsov, D. Bimberg Yu.G. Musikhin and W. Neumann "Tuning Quantum Dot Properties by Activated Phase Separation of an InGa(Al)As Alloy on InAs Stressors" Phys. Rev. B (2000), *in print*.

14. N.N. Ledentsov, M.V. Maximov, D. Bimberg, T. Maka, C.M. Sotomayor Torres, I.V. Kochnev, I.L. Krestnikov, V.M. Lantratov, N.A. Cherkashin, Yu.M. Musikhin and Zh.I. Alferov "1.3 µm luminescence and gain from defect-free InGaAs-GaAs quantum dots grown by metal-organic chemical vapour deposition", *Semicond. Sci. and Technol.* 15 (6), pp. 604-607 (2000).

15. N.N. Ledentsov, M. Grundmann, F. Heinrichsdorff, D. Bimberg, V. M. Ustinov, A. E. Zhukov, M.V. Maximov, Zh.I. Alferov and J. A. Lott "Quantum-Dot Heterostructure Lasers» *IEEE Journal on Selected Topics in Quantum Electronics* 6, 439-451 (2000).

16. D.L. Huffaker, G. Park, Z. Zou, O.B. Shchekin, and D.G. Deppe, "1.3 µm room-temperature GaAs-based quantum dot laser", Appl. Phys. Lett., Vol. 73, pp. 2564-2566 (1998).

17. A.E. Zhukov, A.R. Kovsh, V.M. Ustinov, Yu.M. Shernyakov, S.S. Mikhrin, N.A. Maleev, E.Yu. Kondrat'eva, D.A. Livshots, M.V. Maximov, B.V. Volovik, D.A. Bedarev, Yu.G. Musikhin, N.N. Ledentsov, P.S. Kop'ev and D. Bimberg, "Continuous-wave operation of long-wavelength quantum dot diode laser on a GaAs substrate", *IEEE Photon. Technol. Letters,* Vol. **11**, pp. 1345-1347 (1999).

18. M.V. Maximov, Yu.M. Shernyakov, I.N. Kaiander, D.A. Bedarev, E.Yu. Kondrat'eva, P.S. Kop'ev, A.R. Kovsh, N.A. Maleev, S.S. Mikhrin, A.F. Tsatsul'nikov, V.M. Ustinov, B.V. Volovik, A.E. Zhukov, Zh.I. Alferov, N.N. Ledentsov, D. Bimberg, "Single transverse mode operation of long wavelength (~1.3 µm) quantum dot laser", *Electonics Letters* **35,** pp.2038-2039 (1999).

19. G.T. Liu, A. Stintz, H. Li, K.J. Malloy, and L.F. Lester, "1.25 µm low threshold current density dots-in-a-well (DWELL) lasers", IEEE/LEOS Summer Topical Meeting, Workshop on Nanostructures and Quantum Dots, 26-27 July 1999, San Diego, CA, USA (IEEE Catalog Number: 99TH8455), pp. 19-20.

20. J.A. Lott, N.N. Ledentsov, V.M. Ustinov, N.A. Maleev, A.E. Zhukov, A.R. Kovsh, M.V. Maximov, B.V. Volovik, Zh.I. Alferov, and D. Bimberg, "InAs-

InGaAs quantum dot VCSELs on GaAs substrates emitting at 1.3 µm", *Electronics Lett.*36, 1384-1385 (2000).

21. N.N. Ledentsov, D. Bimberg, V.M. Ustinov, J.A. Lott, and Zh.I. Alferov, "Quantum Dot Lasers: The Promises Come to Reality" Memoirs of The Institute of Scientific and Industrial Research, Osaka, vol. 57, special issue "Advanced Nanoelectronics: Devices, Materials and Computing" (3rd Sanken International Symposium, 14 -15 March 2000) pp. 80-87 (2000).

22. J.A. Lott, N.N. Ledentsov, V.M. Ustinov, N.A. Maleev, A.E. Zhukov, A.R. Kovsh, M.V. Maximov, B.V. Volovik, Zh.I. Alferov, and D. Bimberg "Room Temperature Continuous Wave InAs-InGaAs Quantum Dot VCSEL on GaAs Substrates Emitting at 1.3 µm" LEOS2000, 13-16 November 2000, Puerto Rico, IEEE Annual Meeting Conference Proceedings, Vol.1, pp. 304-305.

23. D. Bimberg, M. Grundmann, N.N. Ledentsov, M.H. Mao, Ch. Ribbat, R. Sellin V.M. Ustinov, A.E. Zhukov, Zh.I. Alferov, J.A. Lott "Novel Infrared Quantum Dot Lasers: Theory and Reality", phys. stat. sol. (a), in print

24. M. Yamada, T. Anan, K. Tokutome, A. Kamei, K. Nishi and S. Sugou "Low-Threshold Operation of 1.3-µm GaAsSb Quantum-Well Lasers Directly Grown on GaAs Substrates" *IEEE Photonics Technology Letters* **12**, pp. 774-776 (2000); K. Nishi, T. Anan, and S: Sugou "GaAsSb quantum well for 1.3 µm VCSEL application" IEEE/LEOS Summer Topical Meeting, Workshop on Nanostructures and Quantum Dots, 26-27 July 1999, San Diego, CA, USA, pp. 39-40.

25. M. Yamada, T. Anan, K. Kurihara, K. Nishi, K. Tokutome, A.Kamei and S. Sugou „Room temperature low-threshold CW operation of 1.23 µm GaAsSb VCSELs on GaAs substrates" *Electronics Letters* **36**, pp.637-638(2000).

26. K. Nakahara, M. Kondow, T. Kitatani, M. C. Larson, and K. Uomi „1.3-µm Continuous-Wave Lasing Operation in GaInNAs Quantum-Well Lasers "IEEE Photonics Technology Letters 10, pp. 487-488 (1998).

27. T. Kitatani, K. Nakahara1, M. Kondow, K. Uomi, and T. Tanaka "A 1.3-µm GaInNAs/GaAs Single-Quantum-Well Laser Diode with a High Characteristic Temperature over 200 K" Jpn. J. Appl. Phys. 39, Part 2, pp. L86-L87 (2000).

28. M. C. Larson, M. Kondow, T. Kitatani, K. Nakahara, K. Tamura, H. Inoue and K. Uomi, "GaInNAs/GaAs long-wavelength vertical-cavity surface-emitting laser diodes", IEEE Photon. Technol. Lett. 10, pp. 188-190 (1998).

29. O.B. Shchekin, G. Park, D.L. Huffaker, and D. Deppe "Discrete energy level separation and the temperature dependence of quantum dot lasers" Appl. Phys. Lett. 77, 466-468 (2000).

30. G. Park, O. B. Shchekin, D. L. Huffaker, and D. G. Deppe "Low-Threshold Oxide-Confined 1.3-µm Quantum-Dot Laser" IEEE Photonics Technology Letters 12, 230-232 (2000).

31. G. Park, O.G. Schekin, S. Csutak, D.L. Huffaker "Room-temperature continuous-wave operation of a single-layered 1.3 µm quantum dot laser" Appl. Phys. Lett. 75, 3267 (1999).

32. D. L. Huffaker, H. Deng, and D. G. Deppe, "1.15-µm Wavelength Oxide-Confined Quantum-Dot Vertical-Cavity Surface-Emitting Laser" IEEE Photon. Technol. Lett. 10, 185-167 (1998).

33. A.Yu. Egorov, D. Bernklau, D. Livshits, V. Ustinov, Zh.I.Alferov and H.Riechert „1.3 µm CW operation of InGaAsN laser" *Inst. Phys. Conf. Ser.* No 166, Chapter 6, IoP, pp. 359-362 (2000).

34. D.A. Livshits, A. Yu. Egorov and H. Riechert „8W convertion efficiency of InGaAsN lasers at 1.3 µm" *Electronics Lett.*36, 1381-1382 (2000)).

35. K. Mukai, Y.Nakata, K.Otsubo, M.Sugawara, N.Yokoyama, H.Ishikawa „1.3 µm CW lasing characteristics of self-assembled InGaAs-GaAs quantum dots" IEEE J. of Quantum. Electron. 36, pp.472-478 (2000).

36. M.R. Gokhale, P.V. Studenkov, J. Wei, S.R. Forrest „Low-threshold current, high-efficiency 1.3 µm wavelength aluminium-free InGaAsN-based quantum well lasers" IEEE Photonics Technol. Letters 12, pp.131-133 (2000).

37. A. Stintz, G. T. Liu, H. Li, L. F. Lester, and K. J. Malloy „Low-Threshold Current Density 1.3-µm InAs Quantum-Dot Lasers with the Dots-in-a-Well (DWELL) Structure" IEEE Photonics Technology Letters 12, pp. 591-593 (2000).

38. O. Blum and J. F. Klem „ Characteristics of GaAsSb Single-Quantum-Well-Lasers Emitting Near 1.3 µm "IEEE Photonics Technology Letters12, pp.771-773 (2000).

39. K.D. Choquette, J.F. Klem, A.J. Fischer, O. Blum, A.A. Allermann, I.J. Fritz, S.R. Kurtz, W.G. Breiland, R. Sieg, K.M. Geib, J.W. Scott and R.L. Naone „Room temperature continuous wave InGaAsN quantum well vertical cavity lasers emitting at 1.3 μm" Electronics Letters 36, pp. 1388-1389 (2000).

40. A.E. Zhukov, A.R. Kovsh, N.A. Maleev, S.S Mikhrin, V.M. Ustinov, A.F. Tsatsul'nikov, M.V. Maximov, B.V. Volovik, D.A. Bedarev, Yu.M. Shernyakov, P.S. Kop'ev, Zh.I. Alferov, N.N. Ledentsov and D. Bimberg, "Long-wavelength lasing from multiply stacked InAs/InGaAs quantum dots on GaAs substrates", *Appl. Phys. Lett.* 75 (13), pp. 1926-1928 (1999).

41. Ch. Ribbat, R. Sellin, M. Grundmann, N.N. Ledentsov and D. Bimberg, *Appl. Phys. Lett. to be published*; F. Heinrichsdorff, Ch. Ribbat, M. Grundmann, and D. Bimberg, "High-power quantum-dot lasers at 1100 nm" *Appl. Phys. Lett.* **76**, 556-558 (2000).